BBC

Science

Higher Level

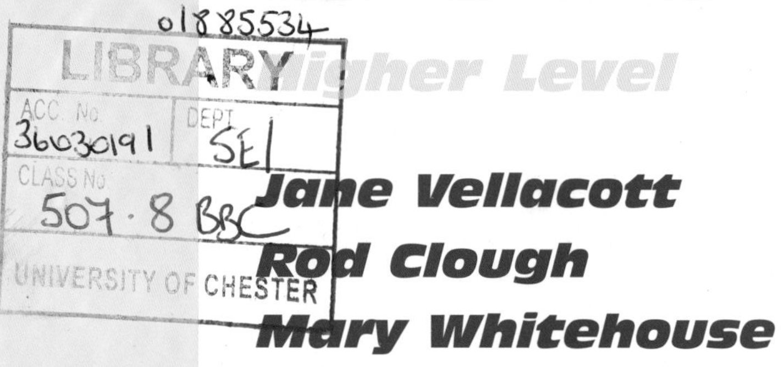

Jane Vellacott
Rod Clough
Mary Whitehouse

Published by BBC Worldwide Limited,
Woodlands, 80 Wood Lane, London W12 0TT

First published 2002
Reprinted July 2003, June 2004

ISBN: 0 563 501294

Colour reproduction by Tien Wah Press Pte Ltd, Singapore
Printed by Tien Wah Press Pte Ltd, Singapore

Acknowledgements
Material from the National Curriculum is Crown copyright and is reproduced by permission of the Controller of the HMSO.
Page 186–187, exam questions based on OCR papers 1794/5 Q8 1998 (Q1), 1787/1 Q2 1995 (Q2),1787/2 Q2 1995 (Q2),1794/6 Q4 2000 (Q3),1794/6 Q6 1999 (Q4),1794/6 Q2 1999 (Q5), material used is © OCR. OCR accepts no responsibility whatsoever for the accuracy or method of working in the answers given.

Contents

PHYSICS

Introduction

About Bitesize

GCSE Bitesize is a revision service designed to help you achieve success at GCSE. There are books, television programmes and a website, each of which provides a separate resource designed to help you get the best results.

TV programmes are available on video through your school, or you can find out transmission times by calling 08700 100 222.

The website can be found at
http://www.bbc.co.uk/schools/gcsebitesize

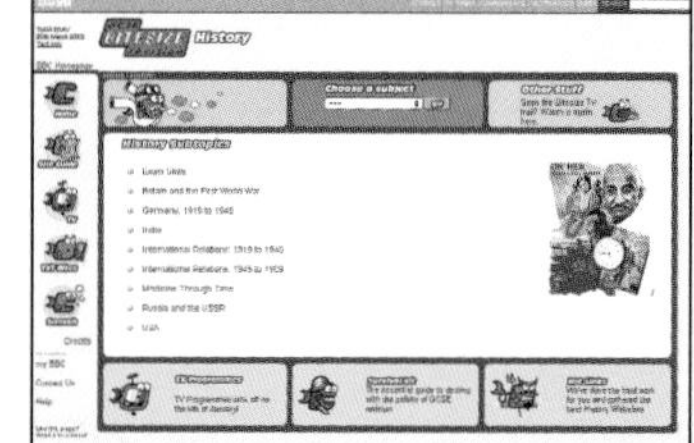

About this book

This book is your all-in-one revision companion for GCSE. It gives you the three things you need for successful revision:

1. **Every topic clearly organised and clearly explained.**
2. **The most important facts and ideas highlighted for quick checking:** in each topic and in the extra sections at the end of this book.
3. **All the practice you need:** in the check questions in the margins, in the practice sections at the end of each topic, and in the exam questions section at the end of this book.

Each topic is organised in the same way:

- **The bare bones** – a summary of the main points, an introduction to the topic, and a good way to check what you know.
- **Key facts** highlighted throughout.
- **Check questions** in the margin – have you understood this bit?
- **Remember tips** in the margin – extra advice on this section of the topic.
- **Exam tips** in red – specific things to bear in mind for the exam.
- **Practice questions** at the end of each topic – a range of questions to check your understanding.

The extra sections at the back of this book will help you check your progress and be confident that you know your stuff.

Exam questions and model answers

- A selection of exam questions with the model answers explained to help you get full marks. There are also some questions for you to try, complete with answers.

Topic checker

- Quick questions in all topic areas.
- As you revise a set of topics, see if you can answer these questions – put ticks or crosses next to them.
- The next time you revise those topics, try the questions again.
- Do this until you've got a column of ticks.

Complete the facts

- Another resource for you to use as you revise: fill in the gaps to complete the facts.

Answers are at the end of the section.

Last-minute learner

- The most important facts in six pages.

How to use this book

This book is divided into the three science subjects, which are sub-divided into units that cover the key GCSE topics of Science. If you have any doubts about which topics you need to cover, ask your teacher.

For many of the units, there are corresponding sections on the video. In such cases, it's a good idea to watch the video sequence(s) *after* reading the relevant pages, but *before* you try to work through or answer the practice questions. This is because the video sequences give you extra information and tips on how to answer exam questions. It's also a good idea to write the time-codes from the video on the relevant page(s) of the book – this will help you find the video sequences quickly, as you go over units again.

The most important and popular sections of the GCSE Science specification (regardless of exam board) are covered by the book – but BITESIZE Science doesn't aim to give total coverage of all topics. So it's important to carry on using your school textbook and your own notes. Because all the main types of GCSE Science questions you will be tested on in the exam are covered, the general tips and suggestions will be useful, even if some of your specific topics do not appear in the BITESIZE Science book. Remember, the skills are transferable to the content of any topic.

Taken together, this book, the BITESIZE *Check and Test* book, and the video cover all the main skills and contain all the core knowledge required in GCSE Science.

GCSE Science

The National Curriculum sets a Programme of Study with four Attainment Targets for Science.

Sc1 Scientific Enquiry is in two parts. **Ideas and evidence in science** considers the power and limitations of science and how different groups have different views about the role of science. There will be questions about these ideas in your written exams. **Investigative skills** are assessed by coursework, which you will have to hand in to your teacher for marking. This is a chance to gain as many marks as possible even before you go into the exam. This book concentrates on ways in which you can improve your marks in the exams.

Sc2 Life processes and living things You might also know this attainment target as Biology.

Sc3 Materials and their properties You might know this as Chemistry.

Sc4 Physical processes This is also called Physics.

Each of these attainment targets is worth 25% of the marks towards your GCSE in Science.

Picking up marks in exams

Follow these tips to make sure you get all the marks you deserve. The examiners cannot read your mind – they can only give you marks for what you actually put down on paper.

Read all the questions carefully, because they contain the clues to the answers.

Make sure you answer the question.

Diagrams – use a pencil to add to a diagram, following the instructions in the question.

Graphs

- Label the axes and show the scale and units you are using.
- Plot each point neatly with a cross.
- Draw a line or curve smoothly with a single line.

Chemical equations

- Write out the word equation first.
- Write down the formula for each substance in the word equation.
- Balance the equation.

Calculations

- You must show working for full marks.
- Write down a word equation to show the ideas you are using.
- Substitute the numbers you know into the equation.
- Work out the answer and show the units.
- Even if you can't do the arithmetic, you may get marks for writing down the equation and units of the answer.

The exam papers

Questions often start with information or diagrams, which will help you understand. Underline important words – they will help you in your answers.

The marks for each part of the question are given. The size of the space and the number of marks give you a clue about how long your answer might be. Some questions ask you to write several sentences to explain your ideas about some science. Marks will be awarded for using good English and the correct technical terms.

GCSE exams

All the GCSE Science specifications cover the four Attainment Targets. There is some variety in the details of the specification. You should find out from your teacher which specification you are taking and obtain a copy.

Linear or modular?

Some GCSE science specifications are **linear**. This means that the exams at the end of the course, in June, will test your knowledge and understanding of all the science you have learned in your GCSE course. There may be three papers: one for each of Biology, Chemistry and Physics; or the papers may be integrated, asking questions about all three subjects on each paper. You need to make sure you know what your papers will be like.

If you are taking a **modular** specification you will have been doing short module tests during the course. The marks from these tests will contribute to the final mark for your GCSE. You will take one or two final exams at the end of the course. This book should help you revise for both the module tests and the final exams.

Single or double award?

Most students take Double Award Science – this means when the results come out you will get a double grade for science – CC or AA for instance. You need to know all the work in this book for Double Award Science. If you are taking Single Award, you should check with your teacher which topics you need to know.

Foundation Tier or Higher Tier?

There are two tiers of examination in Science. This book covers the work for the Higher Tier papers. You must make sure you are entered for the right tier. You should discuss with your teachers and parents which is the best tier to enter. And when you get into the exam make sure you are given the right paper!

Planning your revision

Do you know when your exam is? How long have you got to revise? It's no good leaving revision until the night before the exam. The best way to revise is to break the subject up into BITESIZE chunks. That is why we have broken the GCSE science course into topics. There are 86 science topics to revise in this book. You need to work out how many topics you need to do each day to get it all done before the exam. Use the Contents page to plan your revision. You could write the date by each topic, to show when you will revise it.

Of course, you have other subjects to revise too. It is often better to cover more than one subject in an evening – a change is as good as a rest, so they say. So how about planning all your revision by working out how much time you have before the exams start and then sharing the days out amongst your subjects. Don't forget to leave some time to relax too!

Revision tips

It's no good just reading this book, to learn the material and understand it you need to be active. Here are some ideas to try:

- At the end of each double page, close the book and write down the key facts from those pages.
- When there is a labelled diagram to learn, draw a copy of the diagram without the labels. Look at the labels in the book, close the book, label the diagram and then check how many were correct.
- 'Look, cover, write, check' is a good way of learning all sorts of things – including spellings, equations and formulae.
- Use the glossary to make some flash cards. Write a definition on one side of the card and the word on the other. Look at the word – can you write a definition? Look at the definition. Which word is it? How do you spell it?
- Use flashcards to learn the equations. Include the units for all the quantities in equations.
- Use flashcards to learn the word equations for chemical processes. Write the word equation on one side and the balanced equation with symbols on the other.
- Revise with a friend – flash cards are more fun with a partner.
- Make a set of flash cards to fit in your pocket – great for the day when the bus gets stuck in a traffic jam or you have to wait for a dentist's appointment!

Life processes

- Life processes distinguish living things from non-living things.
- Life processes are carried out by body systems.
- Plant and animal bodies are built from cells, which are often organised as tissues and organs.
- A living thing can be studied at different levels of complexity and organisation – from a single cell to a whole organism.

A Life processes and body systems

KEY FACT

1 Living things are alive because of the life processes (activities) that go on inside them.

Remember
Nutrition in plants happens mainly through photosynthesis.

Life process	What happens in the body
respiration	energy is transfered from food to the organism
feeding	provides nutrition, energy and raw materials for body processes
sensitivity	changes inside and outside the body are detected and the body can respond
movement	animals move from place to place to find the best conditions for life; plants move slowly by growing in a certain direction
reproduction	offspring are produced, to replace dying individuals in the population
growth	a permanent increase in size, due to an increase in the number and size of cells
excretion	getting rid of waste materials, such as urea (in urine), carbon dioxide (a waste product from respiration) or oxygen (a waste from photosynthesis)

2 By adjusting to changes, the body systems keep conditions inside the organism just right for the life processes to occur.

3 Body systems interact so the organism functions as a whole.

Q How might being sensitive help an organism to survive?

B Cells, tissues and organs

KEY FACT

1 Living things are made of cells, which are the smallest units of living matter.

- Microscopic organisms may consist of one cell, but larger plants and animals are made up of many cells.
- Different types of cells carry out different jobs within a living organism.

KEY FACT

2 A group of cells of the same sort make up a tissue. Together, different tissues make up an organ, which has a special structure to perform a particular function.

Q Why can large plants and animals carry out a greater variety of activities than one-celled organisms?

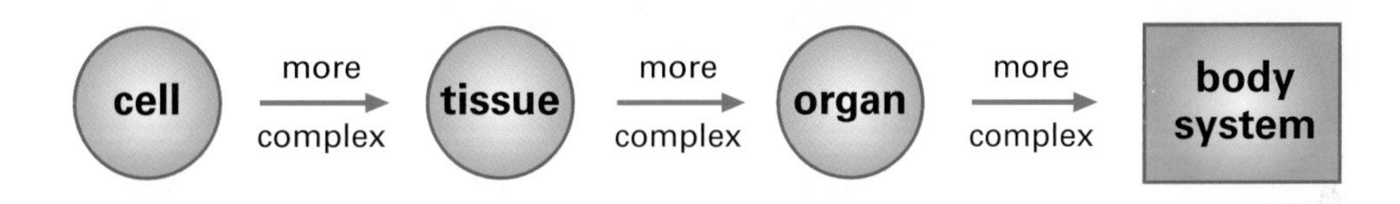

C Case study

As the complexity of an organism increases:

- cells build tissues
- different tissues build organs.

Q Which of the tissues mentioned in the picture on the right are involved in movement?

1 What causes growth?
2 Which life process provides raw materials for building new cells?
3 Why must all living things carry out respiration?
4 Name two examples of human waste products.
5 Place these terms in order of increasing size: nerve tissue, light sensitive cell, eye.

Cells and cell activities

- ➤ All cells have some structures in common, but there are important differences between plant and animal cells.
- ➤ Cells divide during growth and reproduction.
- ➤ The structure of a cell is related to its function.

A Animal and plant cells

KEY FACT

1 Animal and plant cells all share basic features of organisation that allow them to carry out their functions.

- These features include a surface **cell membrane**, contents called **cytoplasm** and a **nucleus**.

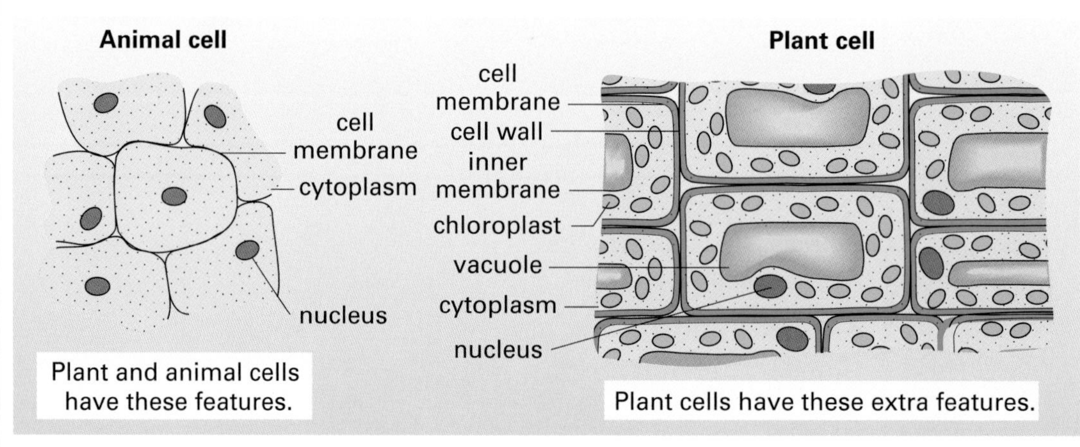

Plant and animal cells have these features.

Plant cells have these extra features.

Remember
Diffusion is a passive movement of molecules down a concentration gradient. Active transport means the cell must expend energy.

2 The cell membrane **controls what moves in and out of the cell.** Only certain kinds of molecules are able to cross the membrane, either passively by **diffusion** or by using an **active transport** mechanism.

3 The **cytoplasm** is **where chemical processes happen**. The rate of these chemical reactions is controlled by **enzymes** [*see page 17*].

- The cytoplasm contains several kinds of smaller structures called **organelles**, each with a particular function.
- **Mitochondria** are the organelles in cells which carry out **respiration** [*see page 28*].

4 The **nucleus** contains the cell's **genetic material** in the form of **DNA**. This genetic material:

- acts as a set of instructions for the cells.
- is passed on from one generation to the next by inheritance.
- gives a living thing its characteristics.

Q Can you describe what organelles e.g. the nucleus, mitochondria, chloroplast do?

KEY FACT

5 Plant cells differ from animal cells in having a cell wall, chloroplasts in the cytoplasm and an inner membrane surrounding a vacuole.

A

- The cell wall is a **rigid support for the cell**, made of **cellulose**.
- The **choroplasts are green** because they contain **chlorophyll** and carry out **photosynthesis** [*see also page 46*].
- The vacuole contains watery sap. Water pressure (**turgor pressure**) in the vacuole pushes the cytoplasm against the cell wall, **helping plants to keep their shape**.

KEY FACT

6 Cells can divide to make more cells, so that animals and plants can grow, repair their tissues and reproduce.

- Ordinary cell division is called **mitosis**. Cell division that happens when sex cells are produced is called **meiosis**.

[*There is more about mitosis and meiosis on page 58*]

B Matching cell structure with function

Cells may look very different even though they share common features. This is because the **structure of each cell relates to its function**.

cell	function and structure
nerve	carries messages around the body; long, thin shape
red blood	absorbs and caries oxygen; has a large surface area of cell membrane
sperm	fertilises the egg cell; has a long tail allowing it to move
root hair	absorbs water; has a large surface area of cell membrane
egg	contains a lot of cytoplasm; if fertilised it develops into an embryo
leaf	carries out photosynthesis; contains many chloroplasts

Q How is the structure of a leaf cell suited to the job it performs?

Example: Explain how the structure of an egg cell is related to its function.

1) describe the **structural feature** 2) say how it helps the **function**

Answer: The egg cell has a lot of cytoplasm because it contains food for a developing embryo.

1 Which feature of plant and animal cells determines what characteristics an individual has?

2 Which two features of plant cells help provide support?

3 a) Suggest a reason why both a red blood cell and a root hair cell have a large surface area.
b) Name three processes by which substances move in and out of cells.

4 Which special feature of a plant cell allows it to photosynthesise its own food?

Body systems

- The human body consists of a number of different body systems – for example, the blood system, breathing system and reproductive system.
- The main body systems of a plant are the shoot system and the root system.

A Human body systems

The body systems **interact** to bring about the healthy functioning of the whole body.

The **nervous system** is composed of sense or receptor cells, which detect changes inside and outside the body. This information is processed by the brain, which causes muscles and glands to respond to change.

The **breathing system** involves the lungs, diaphragm, ribs and rib muscles. Air moves in and out of the lungs when we breathe because of air pressure changes inside the chest cavity.

The **liver** is important as it makes bile, which helps fat digestion, and controls the level of digested food in the blood.

The **blood system** is the major transport system. It is made up of:
- the heart (a pump)
- tubes (blood vessels, such as arteries and veins)
- blood (which carries molecules).

Blood is also important in fighting disease and healing wounds.

The **reproductive system** contains sex organs, which produce sex cells.

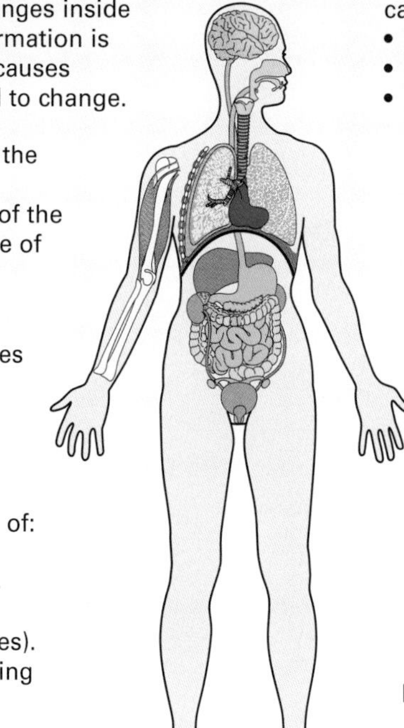

The **skeletal system** is made of bone, cartilage, tendon and muscle. It:
- gives protection to delicate organs
- supports soft tissues
- allows us to move.

The **digestive system** is basically a tube from the mouth to the anus, with different features along its length. Food is broken down (digested) into smaller particles and absorbed into the blood. Undigested food passes out of the body as faeces.

The **urinary system** consists mainly of kidneys, which:
- excrete urea in urine
- balance the amount of water and salts.

Urine is stored in the bladder before passing out of the body.

The **chemical control system** involves glands, which make hormones. Hormones are carried by blood and act on particular target organs. This helps the body respond to changes in body conditions.

Example: Which body systems and processes **supply energy to cells**?

Answer:

- The breathing system – gets oxygen into the body.
- The digestive system – gets food (glucose) into the body.
- The blood system – delivers oxygen and glucose to cells.
- Respiration – transfers energy from food to cells.

Example: The immune system gives protection from disease, mainly provided by white blood cells.

Other body systems are also important:

- skin is a barrier to infection, repairing itself after injury.
- blood clotting helps to seal wounds.
- antibacterial tears bathe the eyes.

Make sure you know the difference between breathing and respiration. Breathing is about getting oxygen from air into the body, and respiration is the transfer of energy from food to cells.

Q Why is a blood system needed in a multi-celled organism?

B The plant body

KEY FACT

1 The main body systems of a plant are the shoot system and the root system.

Remember
Because plants can make their own food by photosynthesis, they are always at the start of a food chain [*see page 72*].

2 The shoot system is made up of the leaves, stems and flowers.

- The main function of leaves is to carry out photosynthesis. Leaves convert simple raw materials from the air and water from the soil into carbohydrate food, which can be stored or built into other more complex substances [*see also page 46*].
- Stems transport materials between the shoot and root system, as well as supporting the shoot system above ground level.
- The flowers are the reproductive structures, where seeds form.

3 The root system absorbs water and minerals, anchors the plant, and may store food.

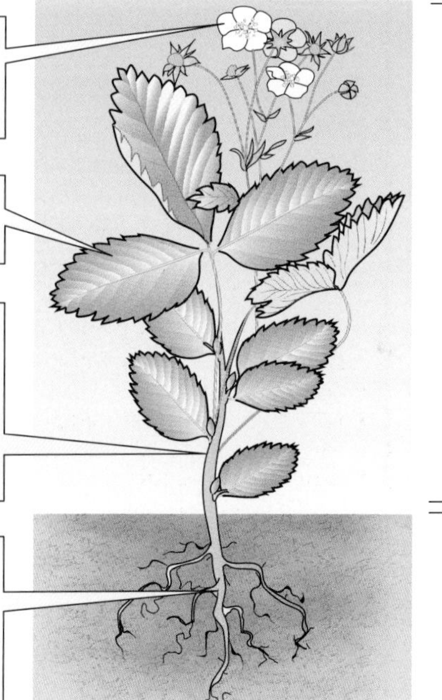

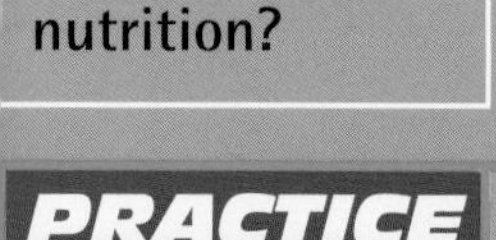
Q Which structure in plants is most concerned with nutrition?

PRACTICE

1 What are the advantages of the stem holding the shoot system well above ground level?

2 Why is the root system important to plant nutrition?

3 Which part of a plant contains the male and female sex cells?

4 Immunity gives protection against infection. Which body system helps give us immunity?

5 Which parts of the body interact to allow it to respond to change in a coordinated way?

Human nutrition

THE BARE BONES

- ➤ Nutrition or nutrients from food provide energy and raw materials.
- ➤ Different nutrients are needed in different amounts.
- ➤ People need a diet that suits their lifestyle.

A The main food types and their uses

KEY FACTS

1 The nutrients in food are needed in order to provide energy for life processes and to provide raw materials for making molecules, cells, and other structures in the body.

2 A balanced human diet includes the correct proportions and amounts of carbohydrates, proteins and fats, as well as minerals, vitamins and water.

food type and chemical elements it contains	how it is used in humans?	items that contain food type
carbohydrates carbon, hydrogen and oxygen a polysaccharide – made of sugar units	• an energy source e.g. sugar • an energy store e.g. glycogen	• sweets, cake and fruit contain sugars *to test: add iodine solution to starch = blue black colour* • potatoes, rice, pasta and bread contain starch *to test: glucose solution and Benedict's reagent are warmed = orange red colour* • vegetables, cereals and fruit contain fibre
proteins carbon, hydrogen and nitrogen a peptide – made of amino acids	• enzymes are protein molecules, which act as catalysts to speed up chemical reactions • cell membranes and organelles are mostly built of protein	• eggs, beans and pulses (e.g. lentils), fish, meat, milk and cheese *to test: Biuret solution mixed with protein = purple colour* • fried foods, fatty meats, oily fish, butter and cheese
fats (lipids) carbon, hydrogen and a little oxygen a lipid molecule – made of fatty acids and glycerol	• cell membranes contain lipid • fat deposits act as insulation • fat deposits cushion the body and prevent damage • a source and storage of energy	• fried foods, fatty meats, oily fish, butter, cheese and vegetable oils

Q Can you explain why a starchy snack (e.g. a baked potato) is more sustaining than eating a sugary snack such as a chocolate bar?

B Vitamins and minerals

KEY FACT

1 A balanced diet also needs to include vitamins and minerals.

- Vitamins are molecules that the body **needs in small amounts to keep healthy**, but **cannot make for itself**.
- Vitamins may be destroyed by cooking, so we need some raw vegetables or fruit.
- Minerals are chemical elements, e.g. iron and calcium, that are essential to health.

2 Example: How the body uses some vitamins and minerals

vitamin or mineral	some uses in body	foods where found
vitamin C	healthy skin and blood vessels	tomatoes, cherries, oranges, lemons
B vitamins	proper working of nervous system	yeast, liver, wholemeal bread
iron	making haemoglobin for red blood cells	meat, spinach
calcium	building strong teeth and bones	milk, eggs

Q Name two sources of iron.

C Diets for a healthy lifestyle

You need a diet that is appropriate to your circumstances

How active you are
Someone doing physical work needs more energy than someone sitting in an office.

Age and gender

child aged 6	7500kJ
girl aged 12-15	9600kJ
woman	9500kJ
man	11500kJ

If you have a medical condition:
high cholesterol: eat low fat foods
diabetic: control sugar intake carefully
high blood pressure: avoid salty foods
kidney not working well: eat less protein

Adults need more energy than children and men need more energy than women due to larger body mass; elderly people are less active than younger people.

Q If men and women are equally active, why do men need to eat more than women?

PRACTICE

1. Which two food types provide most energy in our diet?
2. Why are salads and lightly-cooked vegetables good for you?
3. If someone is actively body building, which food type might they need more of in their diet?
4. What are the four most likely reasons for someone being overweight?

Human digestion

- ➤ The gut (alimentary canal) is a long tube with the main job of processing and absorbing food.
- ➤ Food is broken down into smaller particles as it passes through the gut.
- ➤ Enzymes digest food particles into small, soluble molecules that can be absorbed by the body.

A The structure of the gut

The gut is essentially a **long tube** running from mouth to anus. Different sections of it are **specialised for different purposes**. Nearby organs, such as the **liver** and **pancreas,** also contribute secretions to it.

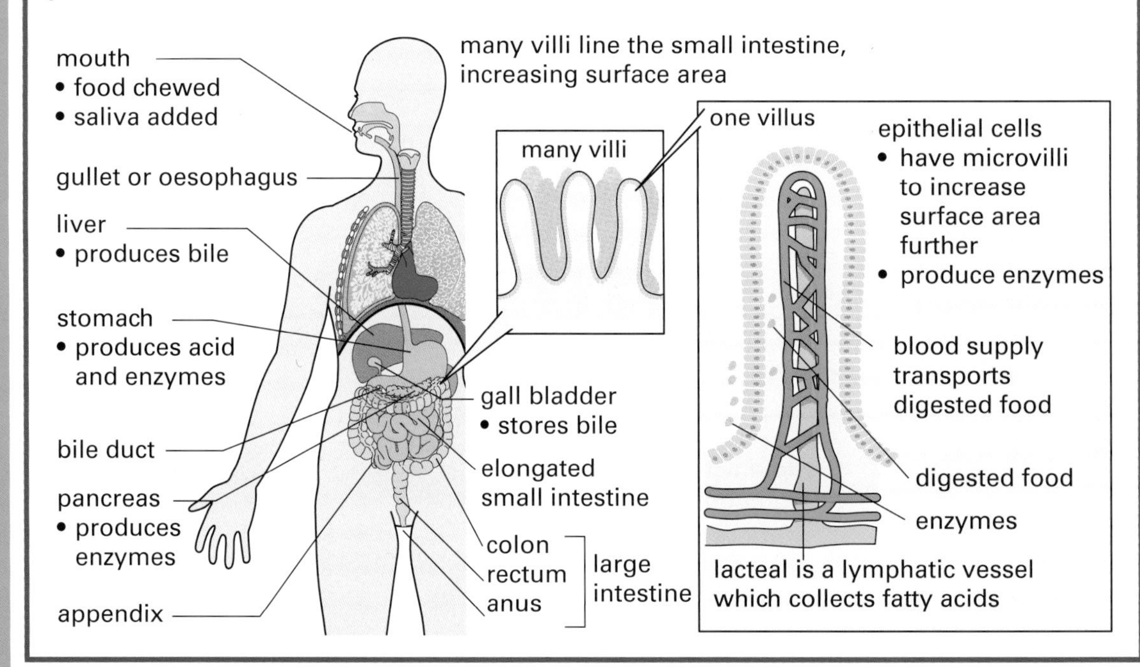

Remember
The small intestine comes before the large intestine, and is responsible for digestion and absorption of digested food.

Q Name an organ (not part of the alimentary canal) that is important in digestion.

B Food processing

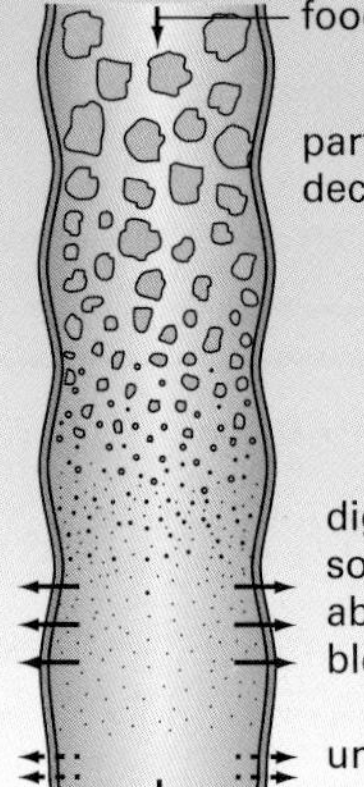

- teeth chop and grind food (physical digestion)
- saliva softens food and adds enzymes
- enzymes help to break down food particles

Peristalsis:
gut wall muscles contract, squeezing food along it

- absorption of digested food
- water reabsorbed

mouth (pH8) produces:
- saliva, containing carbohydrase called amylase, which begins digestion of starch to sugar

stomach (pH2) produces:
- protease digests proteins to amino acids
- acid kills bacteria and helps to break down proteins to amino acids

small intestine (pH9) and the **pancreas** produce:
- carbohydrase
- protease
- lipase digests fat to fatty acids and glycerol

liver adds bile to the small intestine:
- breaks up larger fat droplets
- neutralises acid

Q What is the name for the muscular movement that passes food through the gut?

C About enzymes

KEY FACT

1 Enzymes speed up chemical reactions, and are also called biological catalysts.

- Enzymes are **protein molecules**.
- Many enzymes work inside cells, but digestive enzymes work **outside** them – they are **secreted into the gut**.

KEY FACT

2 Enzymes work by providing a place called an active site where other molecules can bind temporarily.

- Molecules react at the active site, either by **breaking into smaller molecules** or by **joining into larger ones**.
- Only molecules that **match the shape of the active site** can bind on to it.
- Different enzymes have **differently shaped active sites** – that's why only some molecules react with each kind of enzyme.

Remember
Enzyme names sound like the food they digest – add 'ase' to the first part of the food name: carbohydrate is digested by carbohydrase; protein is digested by protease; lipid (fat) is digested by lipase.

3 Enzymes are made of long, thin protein molecules curled up into a ball-like shape. **The exact shape is important**.

- If the shape of an enzyme changes, it cannot join with other molecules and cannot speed up reactions.
- **high temperature** (>50°C) affects the shape, because the protein structure of the enzyme breaks down.
- **change in pH** affects the shape, because pH affects bonds in the protein molecule.

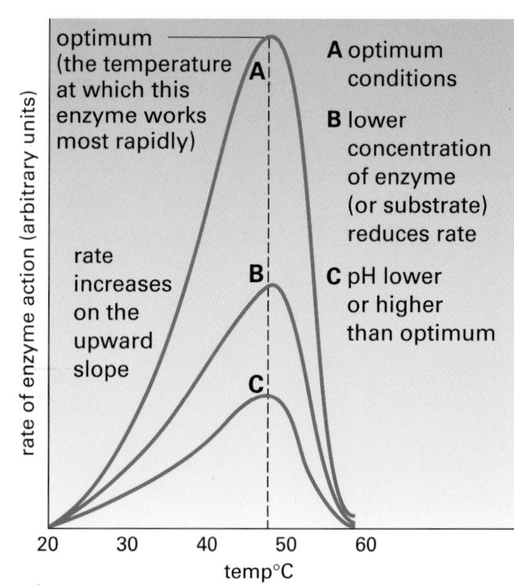

Q Why does amylase in saliva stop working when it reaches the stomach?

You will need to know the general names for digestive enzymes, plus the enzyme 'amylase', which is one example of a carbohydrase.

PRACTICE

1. Name three places where the carbohydrase enzyme called amylase is produced.
2. Describe three ways in which the gut is adapted for maximum surface area.
3. Why do enzymes stop working above 50°C?
4. a) In the graph above, the word 'substrate' means the food that the enzyme is breaking down. Suggest why the line for graph B is lower than A when there are fewer enzyme molecules.
   b) Why is line C lower than line A?

The heart and heart beat

- The heart pumps blood around the body through blood vessels (arteries, veins and capillaries).
- The sequence of events in the heart beat, and the rate of heart beat, are controlled by the nervous system.

A The heart

The heart is mainly made of muscle and beats throughout life, pumping blood around the body.

KEY FACT

The heart is built of four chambers, two atria and two ventricles.

Remember
Atrium = singular, atria = plural. Another word for atrium is auricle.

1. One atrium and one ventricle make up each side of the heart.
2. The left hand side of the heart (LHS) works separately but simultaneously to the right hand side (RHS).
3. The LHS of the heart pumps blood to most of the organs and tissues of the body, and the RHS pumps blood to the lungs.
4. Valves control the direction of blood flow through the heart.

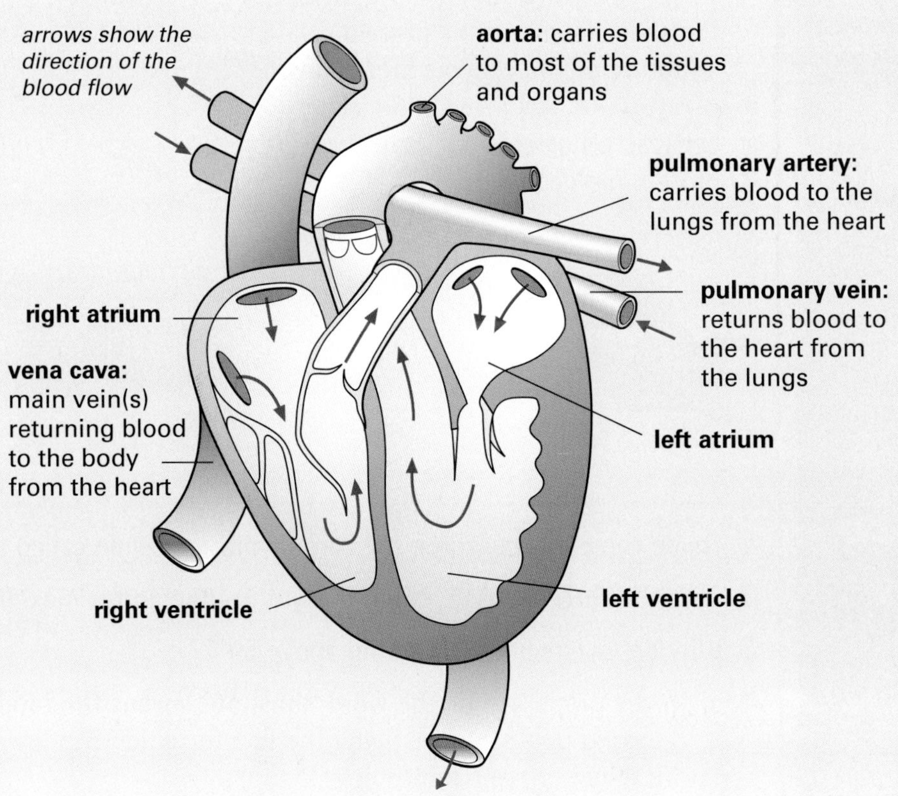

Q Can you name the four chambers of the heart and the largest artery and vein?

B The heart beat

KEY FACT

A heart beat happens when the muscle of the heart contracts in a wave, squeezing blood through the heart and out into the arteries. In-between each contraction the heart muscle relaxes. The heart beat is coordinated by the nervous system.

Remember
The heart rate means the number of heart beats per minute.

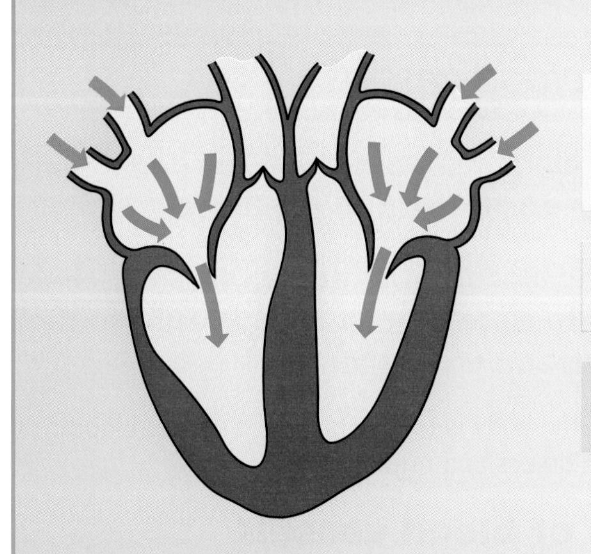

- the atria are relaxed and fill with blood, which enters through the large veins.
- then the atria contract and blood passes into the ventricles.
- valves at the base of the large arteries are closed

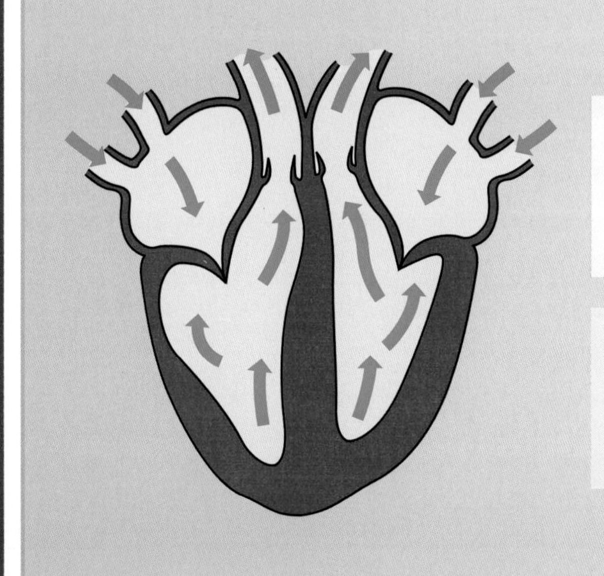

- as the ventricles fill, blood presses against the valves between the atria and the ventricles, closing them
- the ventricles contract strongly, pumping blood into the arteries and pushing open the valves at the base

Q The coronary artery supplies the heart muscle itself. Why is it an important blood vessel?

PRACTICE

1. How is a double circulation achieved?
2. How does the heart muscle bring about a heart beat?
3. Where are the valves in the heart? Why are they necessary and when do they close?

Blood and circulation

- Circulation allows distribution of blood and exchange of materials between blood, tissue fluid and cells, including food, oxygen and waste products.
- Blood is a liquid containing cells, dissolved substances and other components.

A Circulation and blood vessels

KEY FACTS

1 Arteries carry blood to organs from the heart, and veins carry blood from the organs back to the heart.

2 In between, the blood flows through tiny vessels (capillaries) which come into close contact with all cells so that food and waste materials can be exchanged.

Remember
Tissue fluid is the liquid between the cells that make up tissue and organs.

- The largest artery in the body is the aorta, which carries blood from the LHS of the heart to most of the tissues and organs of the body.

The structure of blood vessels

feature	structure	oxygen level in blood	blood pressure	speed and direction of blood flow
artery	thick, muscular and elastic wall with narrow space inside; resists high pressure and pulls back into shape	high	high	fast; blood flows towards organs and away from the heart
capillary	thin wall only one cell thick; tiny in size and passing close to cells; the wall 'leaks' fluid	higher near an artery and lower near a vein		very slow, giving more time for materials to exchange between blood and tissue
vein	thinner wall than an artery with a larger space inside; valves stop backwards flow	lower than an artery	lower than an artery	slower than an artery; blood flows from organs towards heart

Q Why is the circulatory system in humans called a 'double circulation'?

B Blood and its components

KEY FACT

1 The functions of blood include transporting food, oxygen and chemical instructions (hormones – see page 36) to all parts of the body, and taking away waste products. Blood also helps fight against disease (see page 24).

B

KEY FACTS

2 The main components of blood are:
- the liquid part of blood. called plasma.
- red and white blood cells and platelets.

3 The oxygen-carrying molecule haemoglobin is found in red blood cells.

- In the lungs, where there is more oxygen, haemoglobin combines with oxygen to form oxyhaemoglobin.
- Hb + $4O_2$ = HbO_8

  haemoglobin oxygen oxyhaemoglobin
- In the tissues where oxygen is being used up, oxyhaemoglobin splits up to release oxygen to the cells.

Red blood cells flattened, dented discs, which:
- carry oxygen to the cells
- have no nucleus, which gives the dented shape and more surface area to the cell for absorbing oxygen.

Platelets are cell fragments, which are important in blood clotting.

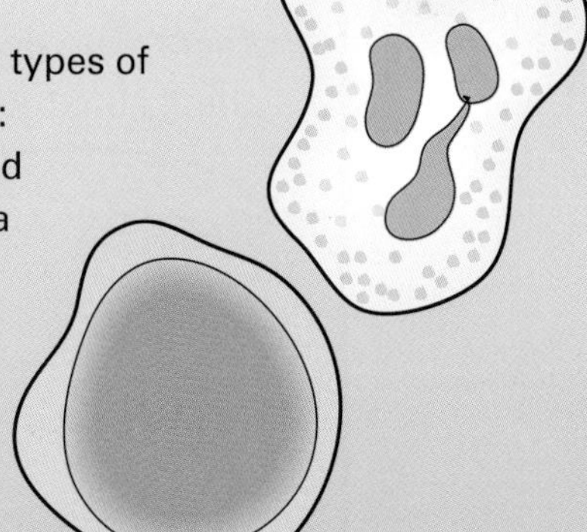

There are several types of **white blood cells**:
- some engulf and destroy bacteria
- some produce antibodies and anti-toxins.

Plasma is the liquid part of the blood, which carries:
- carbon dioxide to the lungs
- digested food
- waste materials such as urea.

Remember
There is a diffusion gradient between the blood and tissue fluid.

Q Why is there a diffusion gradient in the tissues?

PRACTICE

1 Why are there pressure and oxygen level differences in the blood that flows in the aorta and the vena cava?

2 How is the structure of each type of blood vessel related to its function?

3 What makes haemoglobin pick up oxygen, and oxyghaemoglobin split up?

4 There are several types of white blood cells. What are their main functions?

Health and disease

- Causes of poor health include infection, injury, an inherited condition or a lifestyle habit.
- Fungi, bacteria and viruses may cause disease.

A Good and bad health

You need to be able to evaluate information which links the effects of lifestyle to health. This could be data, observations or statements like those below.

1 You are healthy if your body works well, you are coping with life and you feel happy – at least some of the time.

2 The causes of poor health include:

- **infection** – e.g. a cold is caused by a virus, a sore throat is caused by a virus or bacterium, athletes' foot is caused by a fungus.
- **injury** – e.g. the liver might be damaged in an accident.
- **an inherited condition** – e.g. cystic fibrosis, haemophilia.
- a **lifestyle habit** such as drug abuse.
- **mental illness.**

Case study: How can we stay healthy?

There is no doubt that we influence our health by the way we live. What do you think about what these people say about staying healthy?

'My little dog lets me know when it's time for our daily walks, so I keep a lot fitter than many people my age'.

'If I don't eat fibre with my breakfast I get constipated!'

'Breathing in smoke lets harmful chemicals get into blood, and cells in the lungs get damaged.'

' A balanced diet is really important. Let's face it, if you only eat chips and sweets all the time, you'll probably get fat'.

'All drugs such as medicines, caffeine and alcohol, or illegal substances such as heroin, change the way the body works. Drugs can have harmful effects and can be dangerous to health. Of course medicines are controlled, so they are safe to use as instructed'.

"At the office I'm sitting most of the day, so getting to the gym after work is great. It keeps me supple and strengthens my muscles'.

'I'm mad on fruit and veg big time – if I eat plenty I get less spots and feel more energetic.'

Q What type of health condition is each of the following:

- an earache
- a broken leg
- an addiction to tranquillisers
- feeling depressed
- haemophilia?

B Microbes and disease

Microorganisms are of many kinds, and include **bacteria**, **viruses** and **fungi**. Not all microorganisms cause disease – but many do, and they are called **pathogens**.

Bacteria

1μm

- single–celled
- variety of shapes
- microscopic: a light microscope is needed to see them
- genetic material, but no real nucleus
- may be disease–causing (pathogenic)
- feed in a variety of ways

Fungi

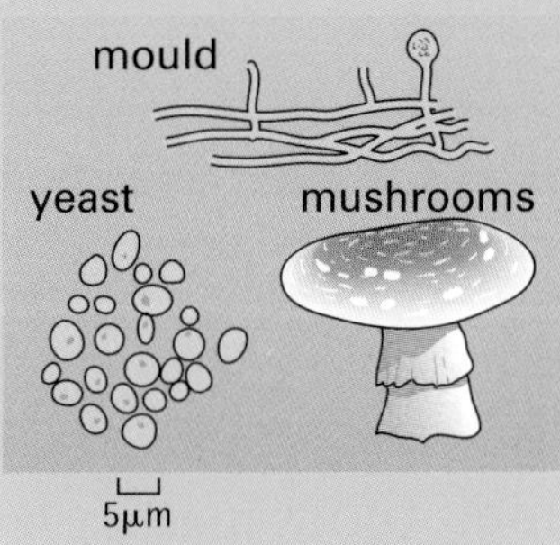

5μm

- mainly many–celled, but yeast are single–celled; moulds are thread–like and form spores; some form mushrooms
- bring about decay by decomposing dead remains of living things
- may be pathogenic
- microscopic, but many may be seen without a microscope

Viruses

- very small, too small to be seen with a light microscope – need to use an electron microscope
- viruses have genetic material, but no real nucleus
- many are pathogenic
- can only live and reproduce inside other cells

$1\mu m = \frac{1}{1000} mm$

Case study: HIV

HIV (short for Human Immunodeficiency Virus) is a virus that enters human cells, including the **white blood cells** which normally protect us against infection. It can be transmitted by sexual activity, or by drug addicts sharing infected needles.

- The virus particles take over white blood cells, using them to make more HIV particles, and **destroying the white blood cells**.
- As yet there is **no 'cure' for HIV** because the virus particles are inside white blood cells, where they are **protected from the body's immune system**.

Q Can you list the three types of microorganisms shown in the diagram in order of size, smallest first?

PRACTICE

1. Measles is caused by a virus. If you have measles, what type of disease is it?
2. Describe three main factors which are important for a healthy lifestyle.
3. Suggest one type of numerical data which would prove a link between smoking and disease.
4. Medicines, alcohol and heroin are all drugs. What does one of the speakers in first section mean when he says that medicines are 'controlled', so safer than other drugs?

Body defences and immunity

- Poor health may be caused by microbes.
- The body's main defences against infection include the skin, the blood system and the immune system.
- Immunisation is a way of protecting against disease.

A The first line of defence

KEY FACT

1 The skin is a barrier to infection. Skin replaces itself continuously, helping minor wounds to heal.

2 The breathing tubes (trachea, bronchi and bronchioles) are lined with cells which have cilia and produce mucus. Dirt and microorganisms are swept away from the lungs in mucus, by the cilia.

3 Saliva and tears contain antibacterial substances that help prevent infection at the eye's surface.

4 Urine contains anti-bacterial substances which help prevent infection via the urethra.

Q What makes skin surfaces vulnerable to entry by pathogens?

B Blood clotting

KEY FACT

Blood clotting stops blood loss, and stops germs from entering the body. A cut blood vessel gets plugged by a clot, which soon turns into a scab. This is the start of wound healing.

Blood clotting and wound healing protect against infection

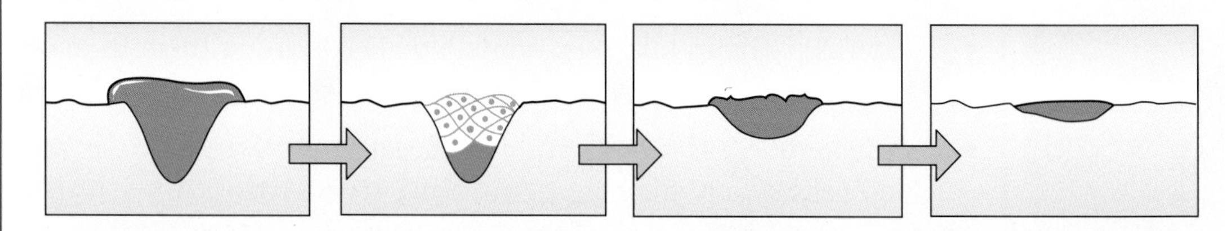

skin is cut; platelets produce clotting factors	clotting factors cause a soluble protein to become insoluble, and a mesh of fibrin threads begin to form a clot, and blood cells get caught in the mesh	the clot begins to dry and becomes a scab	new skin cells grow from beneath the wound and heal it, and the scab dries and drops off

Q Haemophilia is an inherited disease in which the blood lacks clotting factors. Suggest why this can be life-threatening.

C Immune response

KEY FACT

Some white blood cells recognise potentially dangerous particles in the body and fight against them with an immune response.

1. Any **'foreign' protein** entering the body, or a **'non-self' cell such as a bacterium**, is termed an **antigen**.
2. White blood cells can produce **antibodies** which **lock on to antigens**, damaging them or allowing other white cells to engulf them.
3. Some white blood cells can produce **anti-toxins** which neutralise the **toxins** produced by certain bacteria.

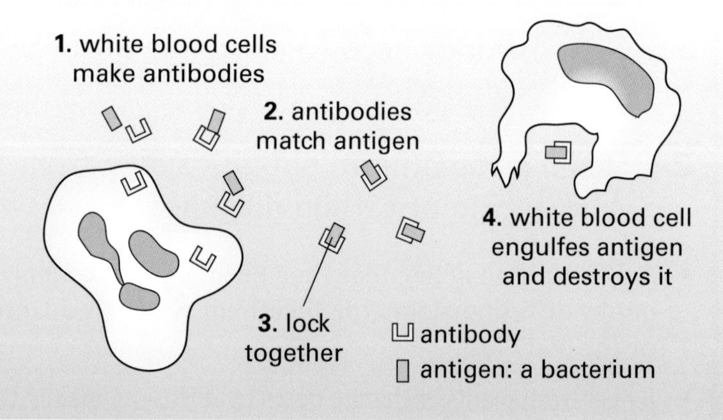

Q Which blood cells are involved in immune response?

D Immunisation

Antigens which are **not dangerous to health** can be **deliberately used** to cause an immune response in a person. This can give **protection against a particular disease** in the future.

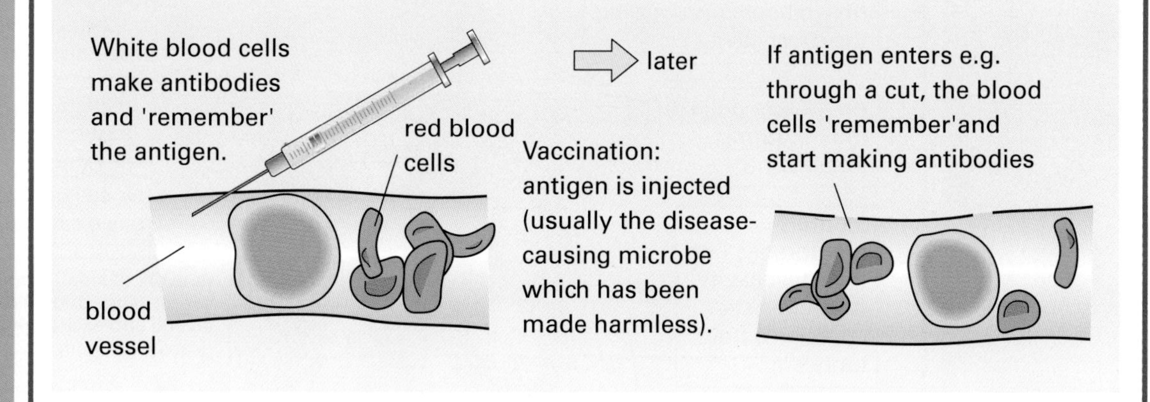

Q How does a vaccination stimulate immunity?

PRACTICE

1. How can our natural defences against measles be improved by medical means?
2. Describe three ways in which white blood cells protect against disease.

The breathing system

- Animals breathe so they can exchange oxygen and carbon dioxide gases with the atmosphere.
- Breathing movements create air pressure differences between the lungs and surrounding air.
- Gas exchange happens in the air sacs (alveoli).
- Smoking damages the breathing system.

A The structure of the breathing system

KEY FACTS

1 Living things need to exchange gases with their surroundings.

2 Animals breathe to obtain oxygen from the air and get rid of carbon dioxide.

- Animals are so active that they need a greater supply of oxygen than they could get by diffusion alone. The breathing system is adapted for getting air in and out of the body fast.
- By contrast, plants exchange gases by diffusion, mainly through pores in the leaf surfaces.

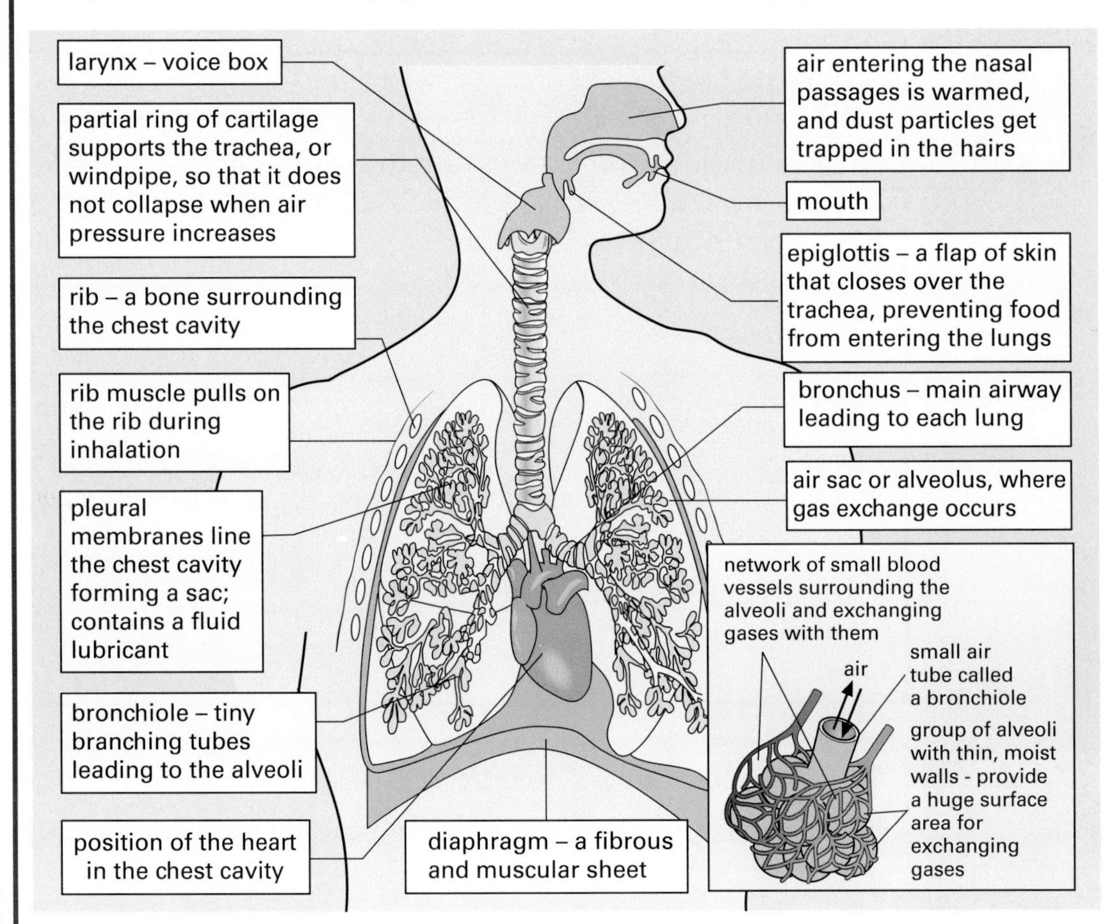

Remember an alveolus is a single air sac (the plural is alveoli).

Q Name the air tubes which lead from the nose and mouth to inside the lungs.

B Breathing movements

1 Breathing is an **active process** which uses energy – muscle movements are needed to make it happen.

- Although you can voluntarily alter the way you breathe for a short time, breathing overall is **controlled automatically by the nervous system.**

2 The direction in which air flows when we breathe depends on the difference in pressure between the inside of the lungs and outside the body.

- If the pressure is **greater inside** the lungs, air **moves out.**
- If the pressure is **lower inside** the lungs, air **moves in.**
- If there is no pressure difference, there is no overall flow.

3

Breathing:	inhalation	exhalation
• **rib muscles**	contract	relax
• **ribs**	up and out	down and in
• **diaphragm**	flattens	domes up
• **volume inside chest and lungs**	increases	decreases
• **air pressure**	drops in chest cavity	rises in chest cavity
• **air flows**	in	out

KEY FACT

Remember
Air pressure is caused by air particles hitting a surface, e.g. the inside of the lungs. Increasing lung volume creates more space and reduces pressure.

Q Why is breathing an active process?

C Smoking and the breathing system

damaging effect	problem it causes
smoke **particles and chemicals** damage the cilia on the cell surfaces and make them less elastic	cilia cannot sweep out dirt and mucus; fluid collects in lungs causing severe breathlessness **(emphysema)**
smoke contains **tar**	smokers at increased risk of **lung cancer**
nicotine in smoke – causes blood vessels to contract	**raises blood pressure** and damages arteries so blood clots may form, and a **stroke** is more likely;
– is addictive	it is hard to give up smoking
carbon monoxide combines with haemoglobin	blood carries less oxygen, so the person does not have as much energy

Q Can you list the main chemicals in smoke which cause problems?

PRACTICE

1 How are the alveoli well-adapted for gas exchange?

2 How do breathing movements bring about air flow?

3 Why do smokers tend to get breathless faster than non-smokers?

Transferring energy by respiration

THE BARE BONES

- Respiration is the transfer of energy from a food source to living cells.
- Aerobic respiration requires oxygen, whereas anaerobic respiration does not.

A Aerobic respiration

1 All animal and plant cells carry out aerobic respiration.

KEY FACT

2 Aerobic respiration extracts much more of the energy present in food than anaerobic respiration does.

- Aerobic respiration occurs in the **mitochondria** of the cell.

KEY FACT

3 During aerobic respiration, chemical reactions happen, which:

- use oxygen and glucose
- break down glucose to produce carbon dioxide and water as waste products
- transfer energy to the cell.

word equation:

glucose + oxygen → carbon dioxide + water (+ energy transferred)

Q Why do cells carry out aerobic respiration?

B Anaerobic respiration

1 During sustained **vigorous exercise**, muscle tissue uses up oxygen very fast, and supply may not be as high as demand. If insufficient oxygen is available, **anaerobic respiration** occurs, and:

- there is **incomplete breakdown** of glucose to a molecule called **lactate** (a waste product, which is used later on in respiration).
- **less energy** is transferred to the cell than during aerobic respiration.

glucose → lactate (+ some energy transferred)

KEY FACT

2 While there is a build-up of lactate in muscle, the muscle cells have an oxygen debt. Later, when vigorous exercise stops, more oxygen becomes available to muscle cells. The cells use the oxygen to fully oxidise lactate to carbon dioxide and water.

3 Other organisms, such as yeast, carry out a type of anaerobic respiration called **fermentation**, in which **ethanol** (alcohol) is one of the waste products.

glucose → carbon dioxide + ethanol (+ some energy transferred)

Q Why do people breathe faster when they exercise?

B Case study: Sports and respiration

Energy continuum describes the types of respiration demanded by different physical activities. Short-term, high-intensity work tends to use anaerobic respiration, while long-term, low-intensity work tends to use aerobic respiration. Many activities are a mixture of anaerobic and aerobic respiration:

physical activity	% aerobic and/or anaerobic respiration	
	aerobic	anaerobic
100m sprint	0	100
100m swim	20	80
boxing	30	70
1500m run/hockey game	50	50
3000m run	80	20
marathon	100	0

Q Suggest a reason why intense activities tend to use anaerobic respiration more than aerobic.

C How energy from respiration is used

Example: A shrew is a small mammal. The graph shows the link between body mass and the rate at which oxygen is used up for different species of shrew.

Energy is used for:

- building up large molecules from small ones
- moving muscles
- keeping a steady body temperature.

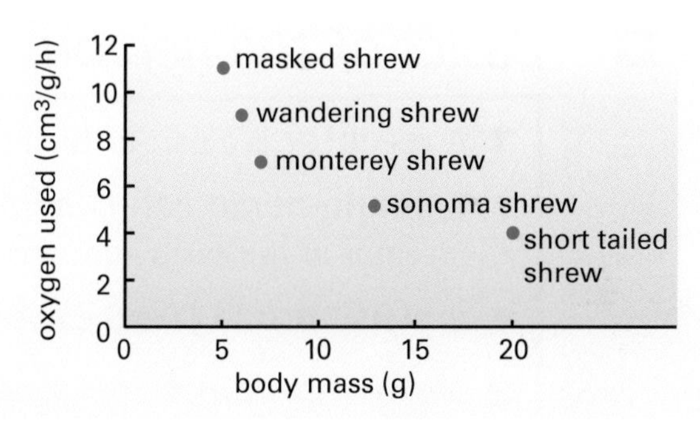

- **A note on units:** $cm^3/g/h$ means cm^3 per gram of body mass per hour.
- **Make sure you can read the graph, e.g.:**
  Q Which species of shrew uses $7cm^3/gram/h$ of oxygen? **A** The unit in the question is on the *y*-axis; read across from the *y*-axis at 7. It is the monterey shrew.
- **Be able to describe a trend, e.g.:**
  Q Describe the relationship between shrew body mass and its oxygen consumption rate. **A** The bigger the shrew, the less oxygen it uses per hour.
- **Know how to interpret data, by saying what it means, e.g:**
  Q Suggest a reason for the trend you described.
  A Bigger shrews lose heat more slowly than smaller ones, so they do not need to carry out respiration as fast, and they use less oxygen.

Questions often ask you to find and use data from graphs.

Q Why do small shrews lose heat to their surroundings faster than a larger shrew does?

PRACTICE

1. Describe three ways in which aerobic and anaerobic respiration are different.
2. In what conditions do muscle cells build up an oxygen debt?
3. Using the graph above, how much more oxygen (in $cm^3/g/h$) does a sonoma shrew use compared to a short tailed shrew?

Sensitivity – detecting change

THE BARE BONES

- To survive, our nervous system must detect and react to stimuli.
- Neurones have varying structures for their different functions.
- The brain acts as a central processing unit for the nervous system.
- Synapses occur at junctions between neurones.

A Survival and sensitivity

KEY FACTS

1 Survival depends on being sensitive to changes inside and outside the body, and responding appropriately.

2 The nervous system provides the body with sensitivity, and helps coordinate life processes.

3 Sensitivity depends on the ability to detect **stimuli**. A stimulus is a condition detected by the body, e.g. temperature of the surroundings, or sugar level in blood.

Q What stimuli are detected when you select a CD and listen to it?

B Basics of the nervous system

1 A **receptor** is a part of the body which detects a stimulus. This could be

- **a single nerve cell** (a **neurone**), such as a pain receptor in the skin, sound receptor in the ear, a light receptor in the eye, or a taste receptor in the tongue.
- **a sense organ** (like the nose) with many smell receptor cells.

KEY FACT

2 The brain and spinal cord make up the central nervous system. The brain is a processing centre for the information detected by receptors. It sends messages to effectors.

3 An effector is the part of the body which **responds** when instructed by the brain, for example, a muscle which moves, or a gland which produces a hormone.

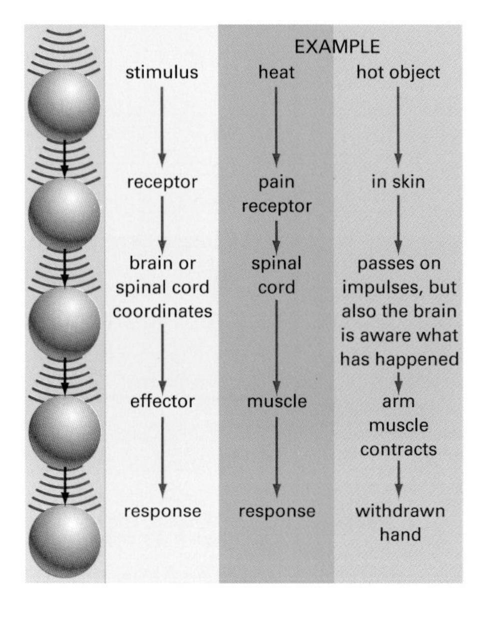

Q Which receptors are stimulated when you send a text message to a friend?

C Synapses

1 A **synapse** is the tiny gap at the junction between two nerve cells, which an impulse must cross.

2 Chemicals called **neurotransmitters** are produced on one side of the synapse. They **diffuse across** the gap, and **initiate the impulse** at the other side of the synapse.

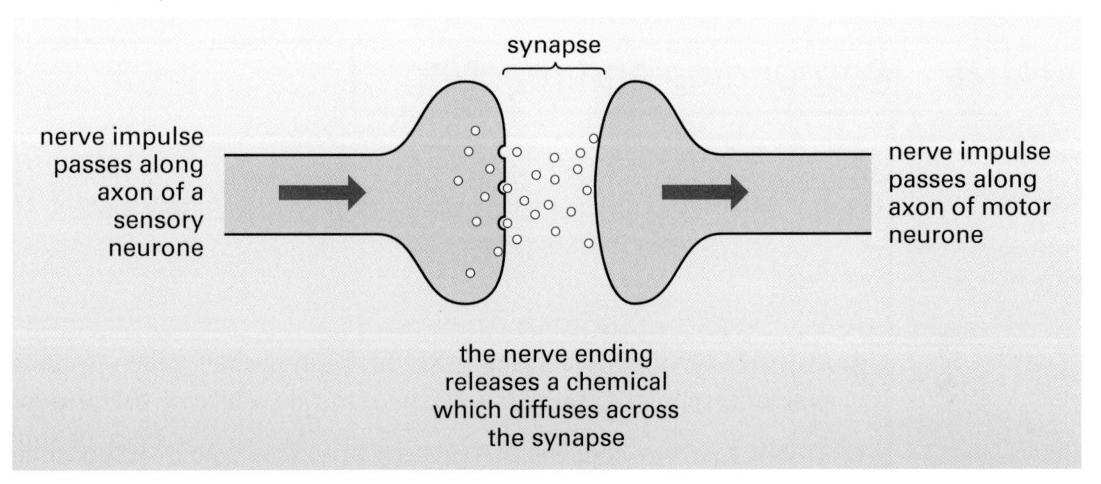

Q Some painkillers have molecules which mimic neurotransmitters, that can block a synapse. Suggest how this causes relief from pain.

D Pain response

Example:

Judy was tired, after a day's shopping, and couldn't wait to sit down with a cup of tea. She sunk into the sofa and kicked off her shoes. Just then she got a text message on her mobile and, as she stood up to get her phone, she experienced a stabbing pain in her foot. Then she remembered her daughter had been sewing on a button the day before . . .

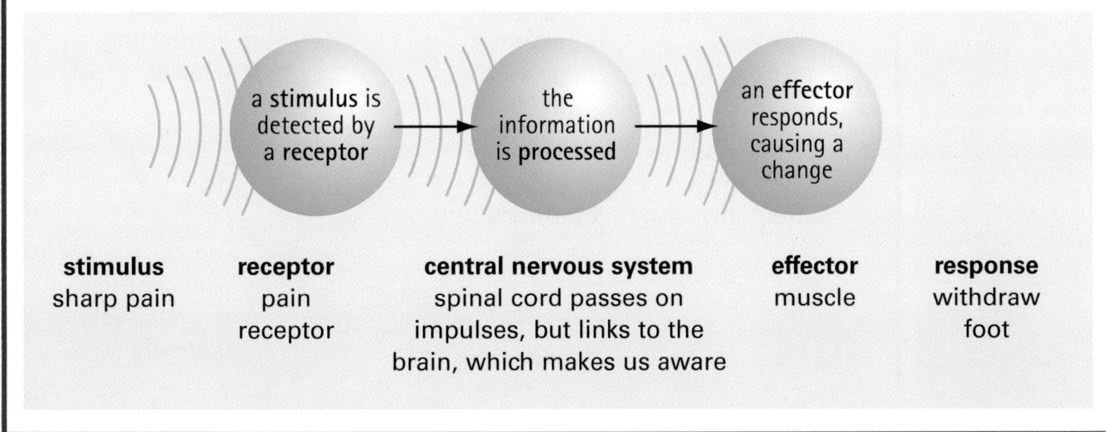

Q In what ways does the central nervous system act as a processing centre?

PRACTICE

1 How is a sensory neurone adapted to its function?

2 What is the job of the fatty sheath around the axon of a neurone?

Sensitivity – response to change

THE BARE BONES

- ➤ Reflex actions happen fast and automatically.
- ➤ Drugs can affect how the nervous system works.

A Responding and reacting

KEY FACT

1 When a receptor is stimulated it sends messages in the form of electrical impulses along a sensory nerve to the central nervous system.

The response may be:

- **voluntary,** or conscious, because the brain decides which response is best. This type of response is learned – e.g. how to ride a bike or perform gymnastics.
- a **reflex**, which happens **automatically**. This type of response does not have to be learned – e.g. we blink when an object comes close to the eye unexpectedly, and the size of the pupil in the eye changes according to the amount of light entering it. The advantage of reflex actions is that they happen **very fast**.

KEY FACT

2 In a simple reflex action, the nerve impulses pass from a receptor along a sensory nerve → to a relay or connecting neurone in the central nervous system → along a motor neurone to an effector, such as a muscle or gland.

Remember
We are aware of reflex actions because of other nerve links to the brain.

Q How do reflex actions help:
a) in everyday activities?
b) survival?

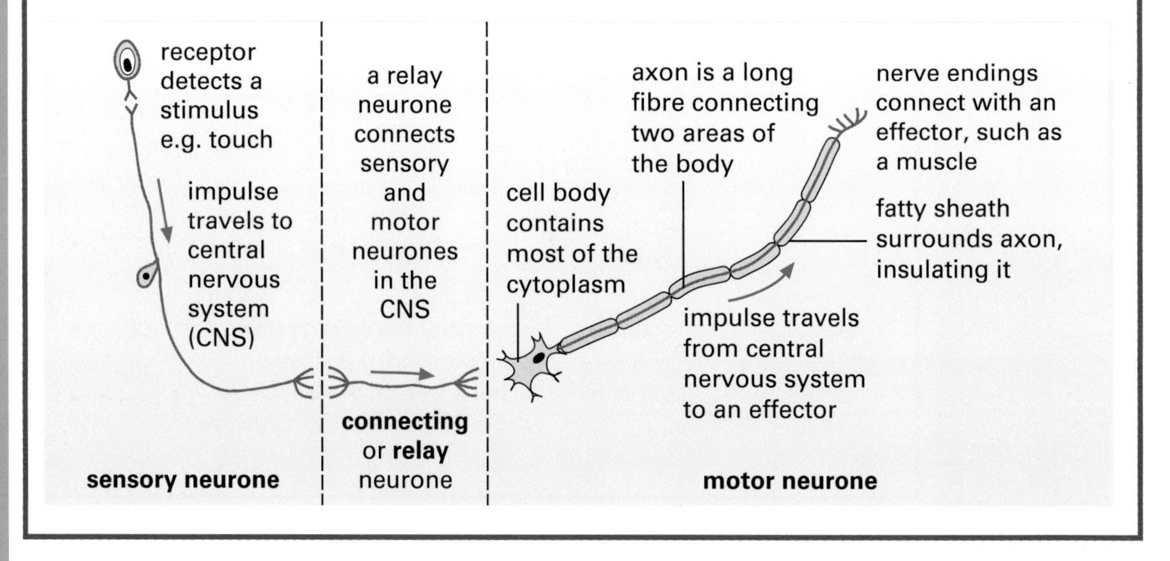

B Drugs and the nervous system

KEY FACT

1 Drugs are found in food materials, tobacco, medicines and substances of abuse.

- Drugs: change how the body works; may affect how we behave; may cause harm.

Examples of common drugs:	
Drug	**Effects**
solvents	affect behaviour; and may damage lungs, liver and brain
alcohol	affects behaviour, may lead to lack of judgement, unconsciousness; slows down reactions; may cause liver damage; can be addictive
nicotine	speeds up the heart rate; causes artery walls to contract, raising blood pressure; can be addictive

2 The abuse or misuse of any drug is potentially dangerous:

- over-use of antibiotics has led to bacteria becoming resistant to antibiotics
- caffeine in coffee is a stimulant which speeds up heart rate and raises blood pressure

3 *Case study:* Quick reactions?

In an experiment to test reaction times, some university students were told to press a buzzer when they heard a bell ring. Their reaction times were measured. Then they drank some beer and their reactions were tested again. Here are the results, showing that their reaction time was affected by drinking beer, and that some students recovered sooner than others.

Student	reaction time/milliseconds				
	before beer	1 hour after beer	2 hours after beer	3 hours after beer	4 hours after beer
Manjit	80	125	95	79	80
Lisa	90	160	100	95	90
Darren	70	150	120	100	90
Jo	85	110	84	85	85

Q Why is it illegal to drive a car once the level of alcohol in the blood has risen above a certain level?

PRACTICE

1 Which of these reactions are likely to be voluntary responses, and which reflexes?
- moving your bare foot off a sharp stone
- 'crying' when you peel onions
- playing a piano
- speaking
- catching onto a rail to save yourself from falling

2 Some poisons block synapses, preventing the transmission of impulses. If a poison affected muscle tissue in this way, suggest a reason why it might be fatal.

The eye and sight

THE BARE BONES

- The eye focuses light into images which the brain can interpret.
- The lens can change shape so that objects near or far can be brought into focus.
- The eye can adjust to bright or dim light.

A The eye

KEY FACT

1 The eye is a sense organ that focuses light rays coming to it from objects in the outside world, so that images are formed which the brain can interpret.

2 The eye is a complex structure and includes features that **help protect it from damage**.

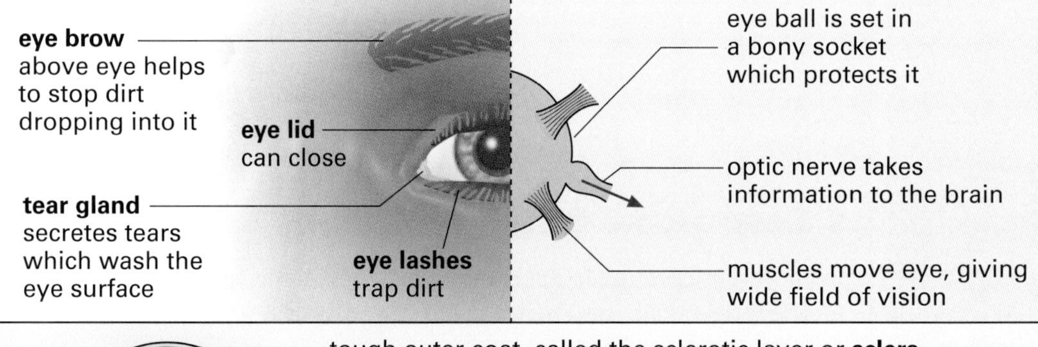

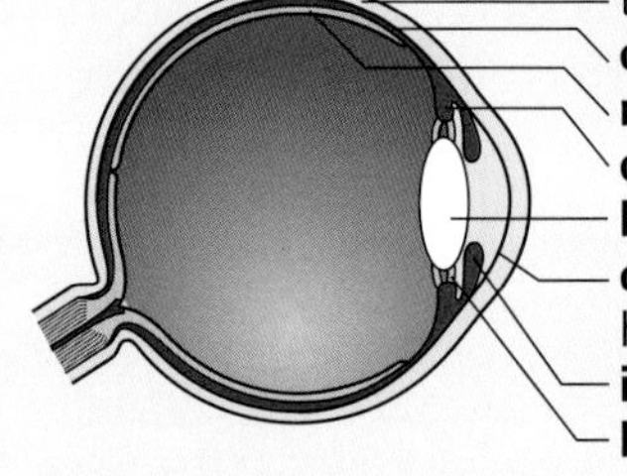

Remember
The axons of sensory neurones collect to form the optic nerve, which passes directly to the brain. It is a short nerve, so we get visual information very fast.

Q In what ways is the eye protected from damage?

B Seeing

1 An **image** is formed when light comes to **focus** on the **retina**.

2 Nerve tissue forms the retina, which is mainly composed of **light-sensitive receptors**.

KEY FACTS

3 The cornea and the lens focus images from near or distant objects.

4 The shape of the lens alters when focusing light from objects at different distances.

B

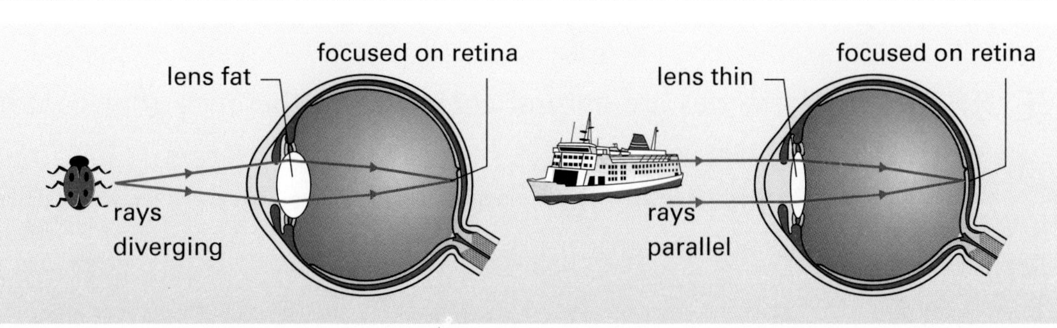

features	looking at close objects	looking at distant objects
light rays reflected by the object	spread apart (diverge) as they enter the eye	enter the eye almost parallel
ciliary muscle	contracted	relaxed
ligaments	slack	pulling on lens
light rays converge	to a greater extent	slightly
image	focused on the retina	focused on the retina

Q Suggest why the lens is made of a flexible material.

C Bright light, dim light

KEY FACT

1 The eye can adapt to dim and bright light conditions.

2 The **iris automatically adjusts** the pupil, to control the amount of light entering the eye. This is an example of a **reflex action**.

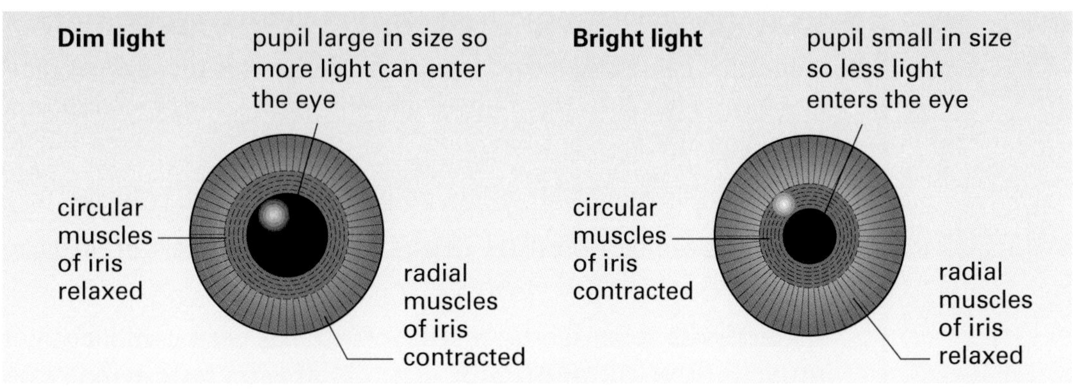

Q What path is taken by impulses in the iris reflex?

PRACTICE

1 Conjuctivitis is an infection of the membrane covering the outer surface of the eye. What natural mechanism is there to combat infection?

2 Which layer in the eye wall is most responsible for transporting food and oxygen?

3 Why is the optic nerve described as a sensory nerve?

4 The image formed on the retina in the graphic appears to be upside down. Why is it that the object looks the right way up to the viewer?

Chemical control in the body

- ➤ Hormones are natural chemicals in the body that coordinate life processes.
- ➤ Glands produce hormones, which are carried in the bloodstream and act on target organs and tissues.

A Glands and hormones

KEY FACTS

1 Hormones are chemicals that the body produces to help coordinate life processes. The hormone system can be called the chemical coordination system.

2 Hormones are produced in the body by glands.

- The body transports hormones in the bloodstream to where they are needed.
- Hormones act on target organs and tissues, bringing about a response or change – for example, insulin causes the liver to convert glucose to glycogen.

3 Hormones may act over a period of time. For example:

- growth hormone adjusts the rate of growth during childhood.
- sex hormones cause egg production by females during much of their adulthood, as part of the menstrual cycle.

4 The effects of hormones can also be immediate. For example:

- adrenaline causes the heart rate to quicken, raises the level of glucose in blood, and diverts blood to the muscles and the brain - thus preparing the body for action.
- insulin has an immediate effect on blood sugar level, reducing it.

5 Hormones can be given as drugs (sometimes as part of medical treatment, sometimes for illegal reasons).

Example 1 In sports, testosterone has been used illegally to increase muscle development. Side effects of extra testosterone can include:

- in males: decrease in sperm production and impotence.
- in females: increase in body hair, voice deepening and irregular periods.

Example 2 Follicle stimulating hormone (FSH) may be used as part of fertility treatment becuase it can bring about maturation of egg cells and their release (ovulation).

A

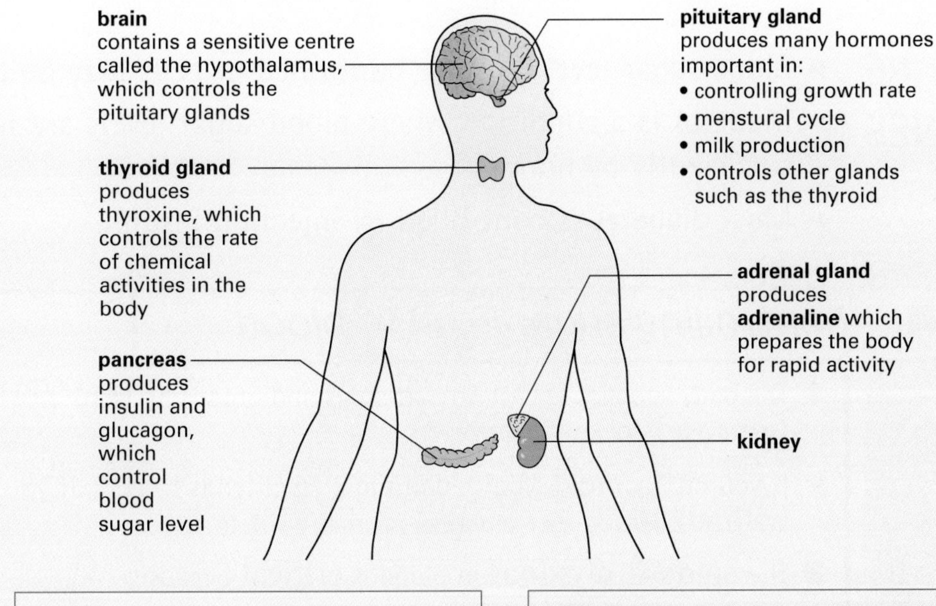

male

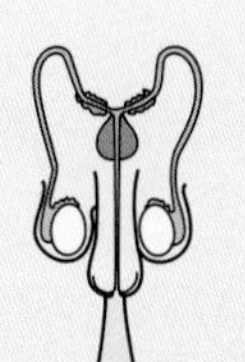

testis
produces testosterone which:
- causes the changes at puberty
- stimulates sperm production

female

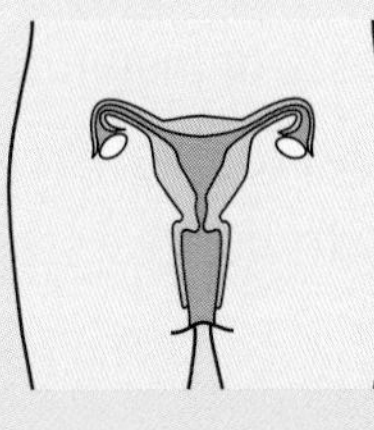

ovary
produces oestrogen which:
- causes changes at puberty
- helps control the menstrual cycle

Comparing coordination systems

nervous system	chemical system
• impulses travel fast / effect is rapid	• hormones transported less rapidly/ long-term effect
• usually short-lived response	• often a long-lasting response
• localised response where effector cells are active	• effects can be widespread as hormone is carried in blood to all tissue; receptor sites may only occur on some cells, so effect is limited.
• only a few neurotransmitters involved	• variety of hormones involved

Q How is the pituitary gland different to other glands?

PRACTICE

1. How does insulin, which is produced by the pancreas, reach the liver, where it mainly acts?
2. Name a target organ for female sex hormones.
3. Describe one effect that hormones cause in males at puberty.

Controlling sugar level in blood

- Blood sugar level must be controlled or cells may be damaged.
- Diabetes is a condition where blood sugar levels are not controlled sufficiently by hormones.
- Most diabetes is controlled by injecting insulin.

A Controlling sugar level in blood

1 The body supplies energy to its tissues and organs by **transporting sugar (glucose) in the blood.**

- If you **eat a meal**, your blood sugar level tends to **rise**. If you **exercise** or **go without food**, your blood sugar may start to **fall**.

2 The **amount of sugar** in blood is **critical** because:

- if blood contains **too much sugar** it draws too much water out of cells by osmosis, damaging them.
- if there is **too little sugar**, organs such as the brain may not get enough energy. (In extreme cases you may become unconscious.)

KEY FACT

3 The blood sugar level is monitored and controlled by the pancreas.

- The pancreas produces two hormones, **insulin** and **glucagon**, with **opposite effects on blood sugar levels.**

What happens when sugar level changes?

If sugar level rises:

pancreas produces more insulin, which causes:

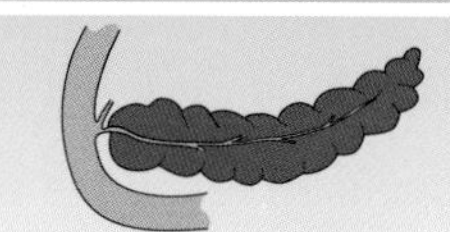

cells to take up more sugar | liver to convert sugar into glycogen and store it.

blood sugar level is lowered

If sugar level falls:

pancreas produces more glucagon, which causes:

cells to take up less sugar (so it stays in blood) | liver to convert glycogen to glucose.

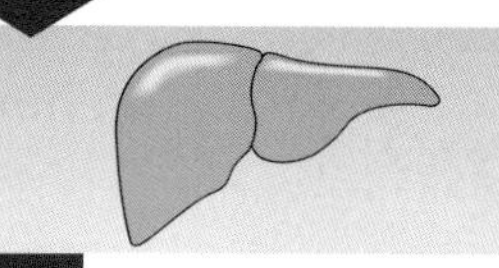

blood sugar level is raised

Remember
Both insulin and glucagon (produced by the pancreas) help in controlling blood glucose level.

Q Can you describe why each of these substances is important in the body: glucose, glucagon, glycogen and insulin?

B Diabetes

KEY FACT

1 Diabetes is a disease where the body does not produce enough insulin (or produces none at all). People with diabetes are called diabetics.

- The most common type of diabetes is inherited. A diabetic lacks the gene which codes for insulin production, so the special cells in the pancreas do not make that hormone.
- Less common is type II diabetes. This starts later in life and is linked to obesity. It can usually be controlled by eating less sugar and controlling other carbohydrate in diet.
- Sometimes diabetes is caused by target tissues not responding to insulin properly.

2 Most diabetes is controlled by injecting insulin directly into the bloodstream.

- Nowadays, insulin is made by bacteria which have been given the gene for making human insulin. This is an example of a genetically engineered medicine.

Q What is the advantage of using genetically-engineered insulin?

C Case study

Nikki is 13, and diabetic. For some years she has managed her diabetes though a programme designed to help her be:

- in control of her condition
- independent
- responsible for staying healthy.

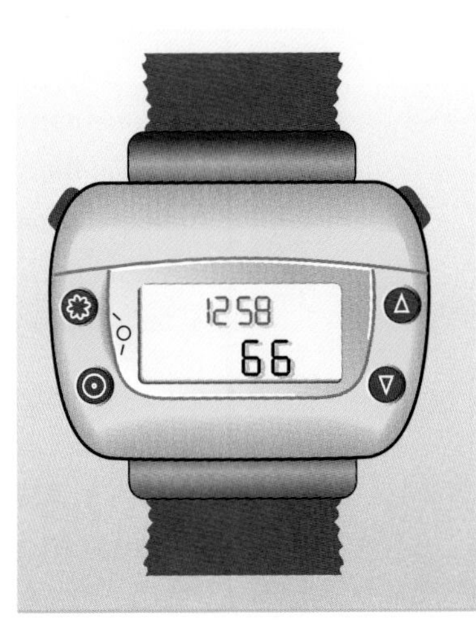

'At first it was a bit scary. But the nurse showed me how to monitor my blood sugar level. Just a pin prick gives me a drop of blood to put in the glucometer and get an instant read-out. I usually do this several times a week and keep a record to show the consultant when I go to the clinic every few months.

I use injections of insulin to keep the glucose level about right, and I'm sensible about eating regularly. I keep some sugar sweets with me in case it drops too low. I don't worry about it now because, if I need to, I can always 'phone the NHS advice line.'

Q What does a glucometer read?

1 A friend of yours spent all his money on sweets and ate all of them in one go. What would be the immediate effect on:
- blood sugar level?
- rate of respiration in the cells?
- insulin production?

2 Suggest two reasons why some people with diabetes have to monitor:
a) their diet b) their blood sugar level.

Sexual maturity

THE BARE BONES

- ➤ Puberty is the time when the sex organs mature.
- ➤ Sex hormones cause the changes during puberty.
- ➤ The menstrual cycle is controlled by hormones.
- ➤ There are various methods of influencing and controlling human fertility.

A Puberty

KEY FACT

1 Puberty is the time in a person's life when the sex organs mature.

2 The body changes which happen during puberty are called the secondary sexual characteristics:

secondary sexual characteristics for boys

puberty happens at around 14-16 years
whole body has a growth spurt
body becomes more muscular
pubic hair grows
beard grows
penis gets larger
testes produce sperms

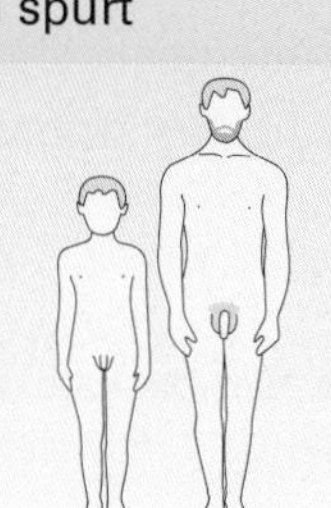

secondary sexual characteristics for girls

puberty happens at around 11-13 years
whole body has a growth spurt
hips widen, buttocks and thighs get fatter, breasts develop
pubic hair grows
ovaries produce eggs
periods start

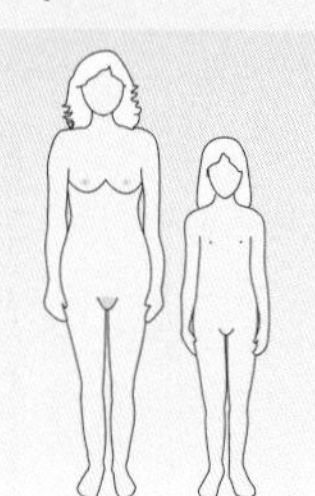

3 Secondary sexual characteristics are brought about by sex hormones:
- testosterone in males
- oestrogen and progesterone in females

Q Can you name and spell the names of the human sex hormones correctly?

B The menstrual cycle

KEY FACTS

1 Having a period is called menstruation. Periods start at puberty and finish at menopause (at around 45-50 years old).

2 Hormones secreted by the pituitary and the ovaries bring about monthly changes:

- ovulation, when an egg is released from an ovary, initiated by follicle stimulating hormone (FSH).
- thickening of the lining of the womb is caused by oestrogen and maintained by progesterone. This means that the womb is ready to receive an egg if is fertilised.

B

3

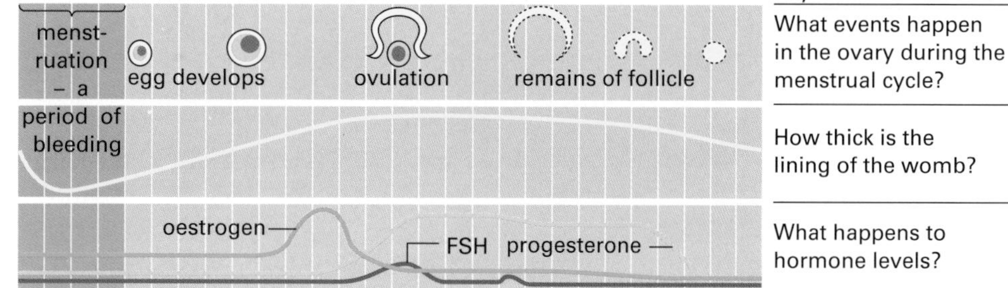

Each menstrual cycle takes 28 days, and is controlled by hormones.

Q Around which day of the menstrual cycle is ovulation most likely to occur?

C Controlling fertility

KEY FACT

1 A reason for low fertility or <u>infertility</u>, is that a woman <u>may not release ova</u> (eggs). <u>FSH</u> can be given as a fertility drug, to <u>stimulate</u> ova to mature and be released from the ovaries.

- there are other causes of infertility which may not be controlled by hormones – for example, infection can block the tubes leading from the ovaries, preventing eggs from reaching the uterus.

KEY FACT

2 Contraception reduces fertility by preventing an egg being fertilised by a sperm.

The main ways of doing this are:

- prevent sperms from coming into contact with eggs e.g. condoms, cap
- stop eggs from being released by the ovaries e.g. the pill
- prevent fertilised eggs implanting in the lining of the womb e.g. IUD, coil
- avoiding sexual intercourse when ovulation is likely to occur.

3 The contraceptive pill contains chemicals very similar to human hormones, such as oestrogen, which inhibits the production of FSH and stops eggs being released by the ovaries. Other effects include:

- making the conditions inside the vagina such that sperm are less likely to survive
- making the vaginal mucus much thicker so it is more difficult for sperm to swim.

Q Suggest a reason why a spermicidal cream (which kills sperms) is often used for additional protection with condoms or caps.

PRACTICE

1 Estimate the number of eggs produced during the lifetime of a woman who starts menstruating at 13 and finishes at 52 years old.

2 Suggest a reason, linked to hormone balance, why a woman might not release ova.

3 In some cultures, people think that using an artificial method of contraception is wrong. Which method of contraception might they still consider using?

Body systems for reproducing

- Male and female reproductive systems are designed to produce sex cells and to allow successful fertilisation.
- The developing baby grows in the mother's womb, nourished by the placenta.
- Controlling fertility provides benefits, but may also bring problems.

A Human reproductive systems

KEY FACT

1 The female reproductive system is designed to produce egg cells, receive sperm from the male, and provide for early development of the offspring.

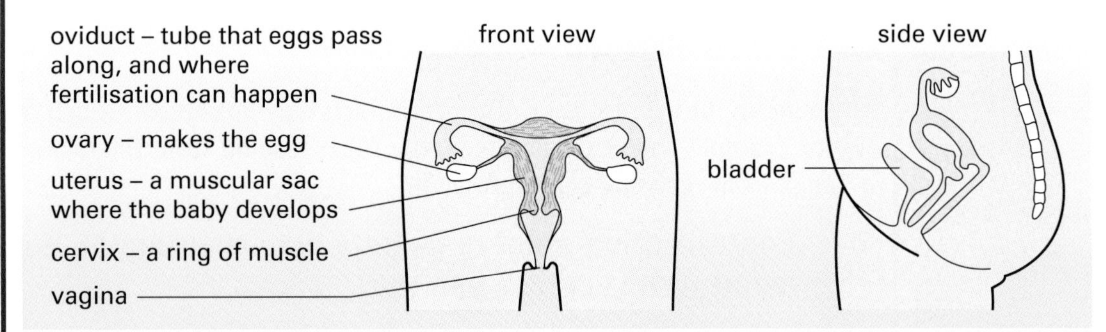

KEY FACT

2 The male reproductive system is designed to produce sperm cells and place them inside the female's body.

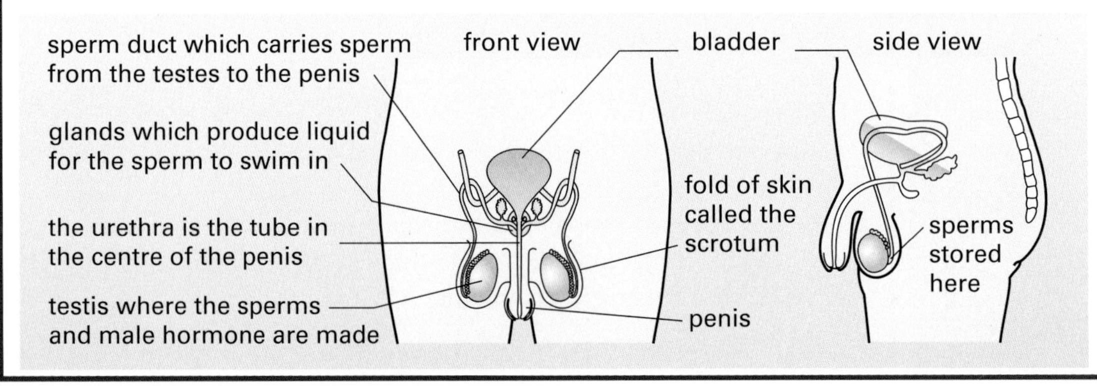

Q Where are sperm cells
a) produced
b) stored?

B Fertilisation and pregnancy

1 The penis is used to place the sperm inside the female's body.

KEY FACT

2 Fertilisation occurs in the oviduct, when the nucleus of the sperm enters the ovum and fuses with the egg nucleus. The new nucleus carries all the information for the development of the new offspring. [See page 59 for more on fertilisation.]

B

KEY FACT

3 The developing baby (first called an embryo, later on a foetus) grows within the womb.

The womb is a good place for a foetus to develop as it is **protected from injury**, it is in a **stable**, sheltered environment and can **expand greatly** to allow growth.

KEY FACT

4 During pregnancy, a special structure called the placenta develops within the womb. This has a rich blood supply which supplies the foetus with nutrition and oxygen.

Q Why does the womb have a rich blood supply?

- The blood supply of the foetus and the mother do not mix directly, although they come into close contact via the placenta.
- Materials, including medicines and drugs of abuse, pass from the placenta to the foetus by diffusion.

Case study: More about controlling fertility

Below, some people talk about their experiences of reproductive technology, the benefits and problems. Try discussing some of the questions with a friend or adult. Why might some people think that no-one should be able to make money out of helping people reproduce?

Why might some people think that no-one should be able to make money out of helping people reproduce?

Q Suggest a reason why in vitro techniques do not always result in pregnancy.

PRACTICE

1 Match the following parts of the female reproductive system with an event that occurs there: **ovary, oviduct, uterus, vagina:**

a a secure place for the foetus to develop
b sperms enter through here
c eggs may be fertilised as they pass along here
d produces eggs.

2 Suggest why humans produce very few eggs compared to fish or frogs.

3 a) Sarah and Tom tried *in vitro* technology to help them start a family. This involves fertilising an egg outside a woman's body and placing it in her womb. Suggest a reason why Sarah felt a failure when *in vitro* fertilisation didn't work.

b) How might years of treatment place a strain on Tom and Sarah's relationship?

c) How might *in vitro* technology affect a couple's relationship if the sperm used is donated by another man?

Keeping the body in balance

- Cells work best if the conditions within them stay fairly constant. This is achieved by homeostasis.
- Balancing the body's needs for water and salts is mainly achieved by the kidneys.
- Body temperature is regulated mainly by sweating, shivering and adjusting the blood supply near the skin surface.

A Homeostasis

KEY FACT

1. The cells of the body need fairly constant conditions to stay alive and work properly. The body's arrangements for keeping these conditions constant are called homeostasis.

2. One reason to keep conditions constant is that the body's **enzymes** are sensitive to changing conditions, and **stop working if the conditions are wrong**.

KEY FACT

3. The skin is a vital organ for homeostasis.

- The skin helps regulate **body temperature**
- It is **waterproof** and protects the body from drying out.
- It protects against **infection**.
- It excretes **urea** in sweat.

Q Why can skin damage from burns over a large area be fatal?

B Balancing water and salts

A small amount of **water is lost** from the body when we **breathe out**, and some is lost through **sweating**. Most excess water is filtered into **urine** by the kidneys.

Q At a party, a child drinks three cola drinks in a row. What would you expect to happen to the amount of urine produced?

C Temperature regulation

KEY FACT

Humans and other large animals have some control over body temperature, whatever changes happen outside. This thermoregulation means that humans can exploit more environments than they could otherwise.

The **hypothalamus** in the brain is the **temperature control centre**:

- **thermoreceptors** detect the blood's temperature as it flows through
- it sends impulses to **effectors** in the skin so that temperature is regulated.

too cold ← Ways of regulating body temperature → too hot	
need to warm the body up	**need to cool the body down**
shivering – involves muscle movement which warms the body up	no shivering
sweating decreases – so less heat is lost as sweat evaporates	**sweating increases** – so more heat is lost as sweat evaporates
blood circulation near skin surface **decreases** so less heat is lost (**vasoconstriction**)	**blood circulation** near skin surface **increases** so more heat is lost (**vasodilation**)
changing behaviour (wrap up warmly, sit by a fire, exercise, have a hot drink)	**changing behaviour** (take off a jumper, sit in the shade, have an iced drink)

Blood vessels in the skin help control body temperature

Q How has thermoregulation helped the success of larger animals?

PRACTICE

1. One Christmas Day a child eats the contents of a whole chocolate selection box.
   a) Which body condition would need to be controlled?
   b) Which body systems are involved in controlling this condition?
2. How do people help keep their water balance stable when they work out in a gym?
3. Which components of skin are involved in thermoregulation?
4. Which tissue acts as an effector when blood vessels change size during vasoconstriction?

Plants and photosynthesis

- ➤ Plants make their own food by the process of photosynthesis.
- ➤ The rate of photosynthesis depends on factors, including light intensity, availability of raw materials, such as carbon dioxide, and temperature.

A What is photosynthesis?

KEY FACT

1 Photosynthesis is the method that plants use to make their own food.

- During photosynthesis, plants use the **Sun's energy** to combine **carbon dioxide** from the air with **water** from the soil to create **glucose** (a **carbohydrate**):

carbon dioxide + water [+ light energy] → glucose + oxygen
(simple raw materials) (carbohydrate food) (by-product)

KEY FACT

2 Most photosynthesis takes place in leaves. Chloroplasts in the leaf cells contain the green pigment chlorophyll, which absorbs light energy, making it available for photosynthesis.

Remember
Photosynthesis and respiration happen in plants.

Q Why is photosynthesis described as a 'building up' process?

3 Plants use glucose to provide **energy** and **raw materials**. Glucose can be:

- built into **starch**, which is a food store, or into **cellulose** for cell walls
- combined with other nutrients to make **amino acids** (building blocks of proteins)
- built into **lipids**, which can also be a food store – e.g. sunflower oil.

4 Although plants use light energy in photosynthesis, they also release energy by **respiration**, just as animals do.

- photosynthesis only happens when there is sufficient light.
- respiration happens all day and night.
- In **daytime**, the balance of photosynthesis and respiration means **more oxygen is given out** than is taken in. At night, more carbon dioxide is given out.

B Investigating the rate of photosynthesis

This is one way of estimating the rate of photosynthesis. Here, the rate of oxygen production shows how fast photosynthesis is happening – measured by counting the number of oxygen bubbles that appear in a certain time. You could collect the gas and measure its volume accurately.

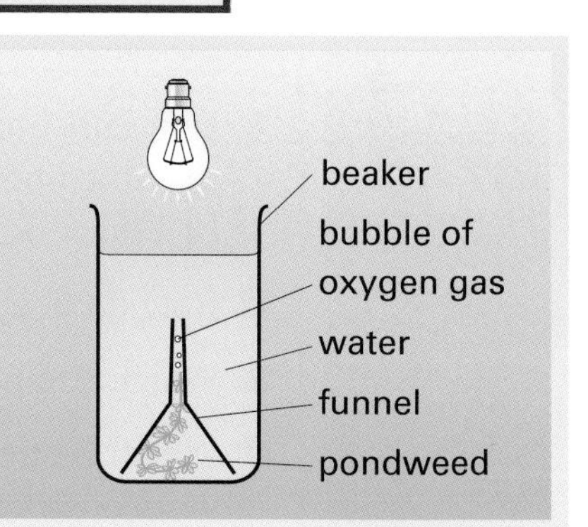

B

Several factors can **limit the rate of photosynthesis**:

- amount of chlorophyll
- light availability
- amount of carbon dioxide
- temperature

Example: Plant growers can use artificial lights and add extra carbon dioxide to air inside a greenhouse to increase the rate of photosynthesis and make plants grow faster. How might plant growers check that adding extra carbon dioxide does increase the yield of a greenhouse crop?

Answer: The grower could try one greenhouse with extra carbon dioxide and one without.

To do a proper test, the grower would have to:

- vary the factors being tested – e.g. the concentration of carbon dioxide.
- keep other factors constant – e.g. type of plants used, the temperature, the amount of watering.

The result of an experiment might look like this:

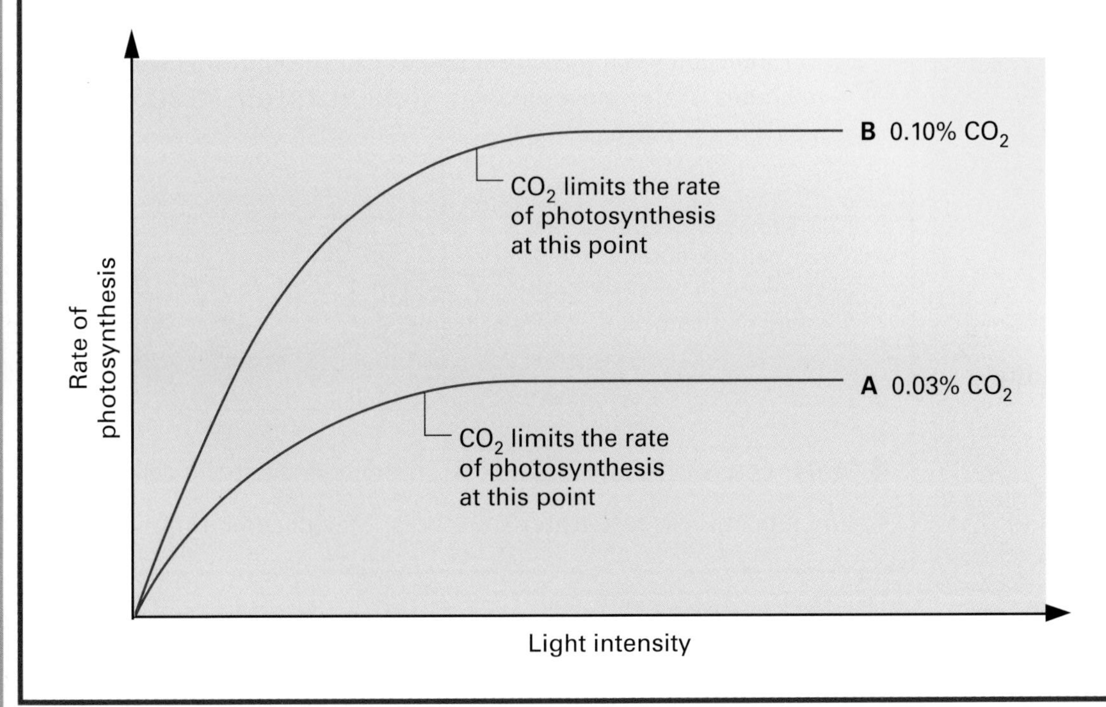

Q If the concentration of carbon dioxide shown in the graph (right) was 0.08%, where would you expect the line of the graph to be?

PRACTICE

1. a) How do the raw materials for photosynthesis reach plant cells?
   b) How are assimilated products transported around plants?
2. Which areas of the plant would be using sugar most actively, and why?
3. Using the graph above, what happens to the shape when the rate of photosynthesis is increasing?
4. How can you tell from the graph that the concentration of CO_2 was a limiting factor?
5. Magnesium is needed for a plant to make chlorophyll. Why might a lack of magnesium in the soil limit the rate of photosynthesis?

Water and transport in plants

THE BARE BONES

- Plants use water for: support, transport and for photosynthesis.
- Transpiration is the loss of water from a plant surface (e.g. a leaf).
- Minerals are important for healthy plant growth.
- Plants have two networks of tubes, xylem and phloem, that transport materials from one part of the plant to another.

A Water and plants

KEY FACT

1 Plants use water for support, to transport minerals from their roots or nutrients from their leaves, and for photosynthesis. Water is also an essential part of all plant cells.

2 In transporting water, plants make use of **osmosis**. Osmosis is the **diffusion of water molecules between solutions of different concentrations**.

3 Water molecules can pass from cell to cell through the **semi-permeable** surface membranes. Water moves along a **concentration gradient**.

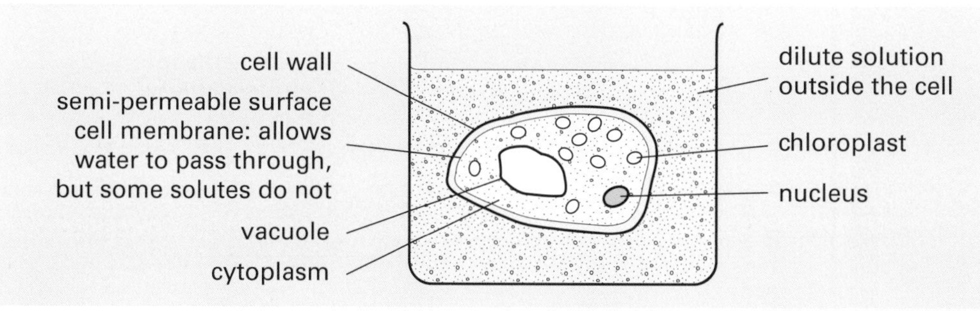

4 Water enters cells and pushes the contents against the cell wall, **giving support**.

- When plant cells lose water from cells, they become floppy and a plant may wilt.

Q What is the difference between a solute and a solvent?

B Transpiration

1 Transpiration is the evaporation of water from plants, especially from their leaves. Although plants can't afford to lose too much water, transpiration has advantages – it cools the leaves, and it helps to draw mineral-containing water up from the roots.

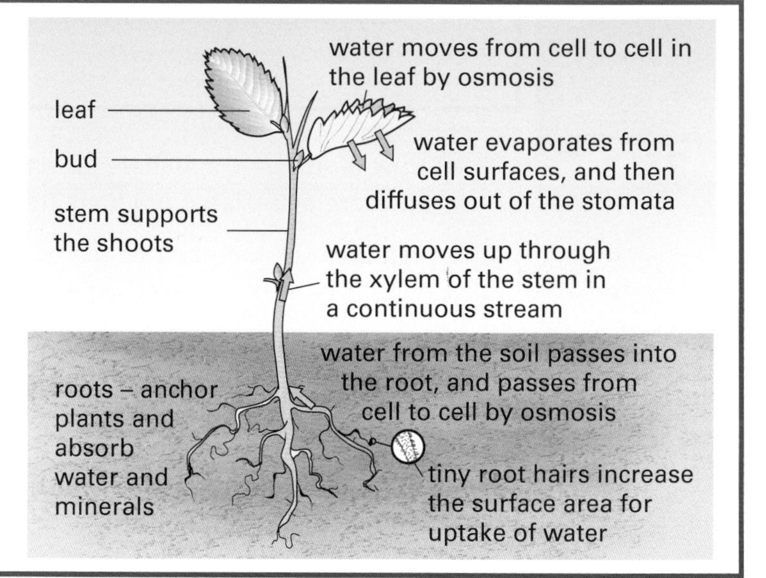

Remember
Plants need stomata because they have to exchange gases with their surroundings. As a result, water vapour is lost from leaves, too.

B

What conditions affect the rate of transpiration?

condition	reason why it affects transpiration
if the **temperature increases**	water molecules evaporate from cell surfaces faster
it is a **windy day**	moving air takes water vapour away from the leaf surface, increasing the diffusion gradient
the **air** around the leaf **is dry**	water evaporates from cells more easily
the number of **stomata**	the more stomata, the faster transpiration happens
a waterproof **waxy cuticle**	prevents water moving through
the amount of **light** (because stomata close in the dark)	no gas exchange can occur

Q What weather conditions would slow down the rate of transpiration?

C Water, minerals and nutrients

KEY FACT

1 Minerals are various chemical elements needed by plants (or compounds that contain these) that are found naturally in most soils. Minerals are needed for plants to grow healthily.

Minerals needed are:

- **nitrate** for making proteins so there is good leaf growth
- **potassium** and **phosphate** for chemical processes, and to promote good bud and root growth

Q What is the difference between organic and inorganic fertilisers?

2 Minerals are also found in **fertilisers**. **Organic fertilisers**, such as manure and compost, come directly from the remains of living things. **Inorganic fertilisers**, which usually contain nitrogen, potassium and phosphate, are made by various industrial processes.

KEY FACT

3 Two separate systems of tube-like vessels make up the transport or vascular tissues of a plant: xylem carries water and phloem carries dissolved sugar.

PRACTICE

1. Why does a plant lose water from its leaves?
2. How does water move along a concentration gradient?
3. How does water move out of the stomata on the leaf surface?
4. A student took two leaves of the same type and size. He covered the lower side of leaf A with grease, and the upper side of leaf B with grease. He weighed both leaves, then left them by a sunny window, and reweighed them four hours later. Here are the results:
   Leaf A lost 2% of its mass, and leaf B lost 10% of its mass. Explain why there was a difference in the amount of water lost by leaf A and leaf B.

Controlling plant growth

➤ Plants make growth movements in response to light, moisture, and gravity.

➤ Hormones control how plants grow.

A Control of plant growth

KEY FACT

1 Plants respond to light, moisture and gravity by growing in a particular direction. These growth movements are called tropisms.

- Shoots grow towards light and away from gravity.
- Roots grow towards moisture and gravity.
- The advantage is that whichever way a seed lands in the soil, its shoot will grow upwards and its root will grow downwards.

KEY FACT

2 Hormones are chemicals that coordinate plant growth. They work at very low concentrations.

Hormones influence:

- how fast cells divide
- the elongation of cells at the tips of roots and shoots, speeding it up or slowing it down.

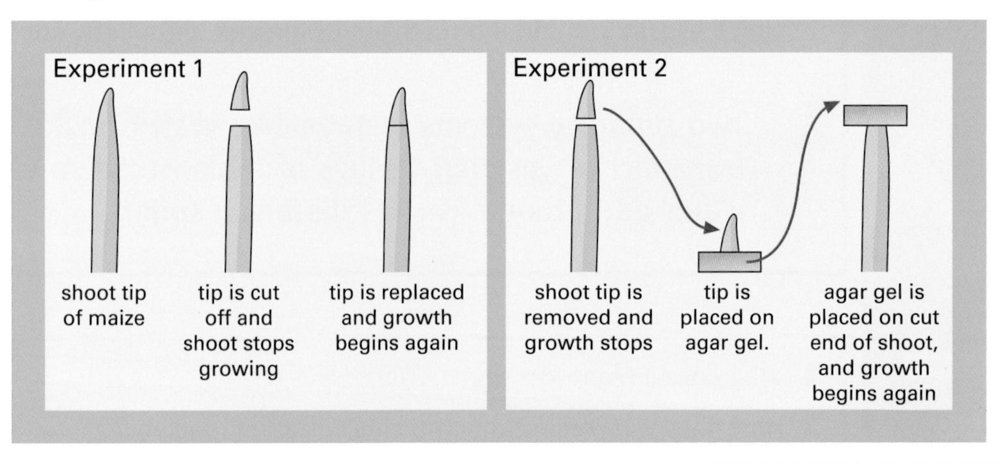

Q What are tropisms?

B Auxins

1 The hormones that have the most widespread effects in plants are called **auxins**.

2 **Auxins** can cause **uneven growth** on either side of a shoot or root tip, so that one side grows faster.

- In **shoots**, auxin **increases** the rate of cell **growth**.
- In **roots**, the auxin **slows** the rate of cell **growth**.

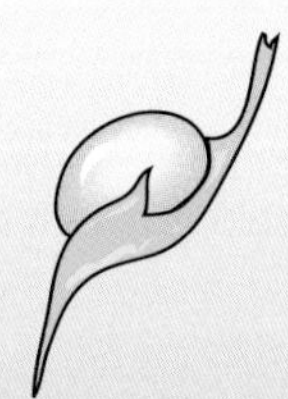

3 The growth of a plant towards light is called **phototropism**.

- Auxin is involved in **phototropism**.
- There is more auxin on the side of the shoot away from light, so it grows faster and curves towards light.

1 shoot in even light grows straight

2 shoot in darkness grows straight

3 shoot in one-sided light, bends towards light

4 shoot in one-sided light with tip covered, grows straight

Q How do hormones help plants get maximum light in a shady woodland?

1 How does gravity affect plant shoots and roots?

2 What is auxin? How might scientists find out if there is auxin in a shoot tip?

3 How does auxin affect cell growth in a shoot?

Plant hormones in action

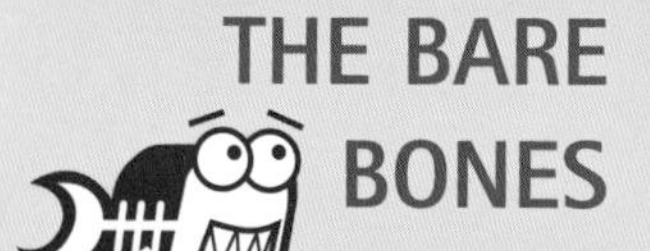

THE BARE BONES

- Hormones are used in horticulture and agriculture.
- Auxin is a hormone, which affects plant growth.

A How do plant hormones work?

There have been many investigations into plant growth, giving much evidence. Even so, piecing it all together to get a proven theory is a bit like doing a jigsaw with some pieces missing.

KEY FACT

1 Scientific theories can take a long time to develop as evidence is collected from different sources.

Example: Why do plants grow towards light? One theory was that light destroyed auxin, giving a greater concentration on the shaded side of a shoot. Look at how further experiments gave more information:

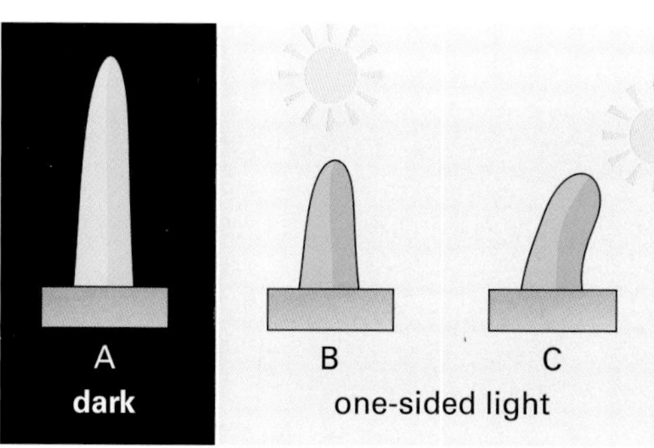

Shoot tips are placed on agar blocks A, B, and C.

The total amount of auxin in each block is analysed, and is about the same.

More auxin is found on the shady side of block C.

KEY FACT

2 Auxin may be actively transported within a stem.

Scientists now believe that auxin is **actively transported** from one side of a shoot to the other. As more and more information is collected, a scientific theory starts to develop.

There are many other growth promoters and inhibitors in plants, which bring about **longer term** growth patterns, such as seasonal growth, leaf fall and fruit ripening.

KEY FACT

3 Plants grow more rapidly in the dark than in the light.

- Plants needs to elongate stems to reach light. In good illumination, more energy goes into building leaf cells so photosynthesis can happen.
- gardeners 'force' the growth of stems, such as rhubarb, for a better crop.

Q Why can it take many years for a scientific theory to develop?

B Plant science in business

KEY FACT

1 Auxins can be used to <u>generate many plants from cuttings</u>, regulate <u>fruit ripening</u> during transport, and even <u>kill weeds</u>.

Using auxin

procedure	effect	how it works
cut the bud off the main shoot	more side shoots develop	auxin from the main bud inhibits growth of side shoots
spray fruit with auxin	more fruit sets	increases pollination, fertilisation and fruit formation
dip shoot cuttings in auxin-containing powder	roots grow faster from the cut shoot	auxin stimulates the growth of small roots (although it can inhibit larger side roots)
spray auxin-containing weedkiller	weeds die	auxin disturbs cell chemistry and causes excessive growth

Q Auxin weedkillers are absorbed more rapidly by broad-leaved plants than grasses. Why does this make them suitable for use on lawns?

KEY FACT

2 <u>Ethene</u> (ethylene) is produced in low concentrations by plants, and causes <u>ripening of fruits</u>. Unripe fruits are transported because they are less likely to be damaged or rot. Just before arrival at supermarkets, ethene is used to bring on ripening.

PRACTICE

1 Why do scientists believe that auxin moves to the more shady side of a shoot tip?

2 The classic theory for explaining how gravity affects roots, proposes that:
- auxin collects on the lower side of a root due to gravity
- growth slows on the lower side
- the root curves downwards.

Which facts in this theory are known to be true?

3 Suggest why a plant grower might want to prune the main shoot of a bush.

Variation and genetics

THE BARE BONES

- Variation can be caused both by genes and by the environment.
- DNA acts as a code for inherited characteristics.
- The Human Genome Project has the potential to lead to cures for many genetic diseases.

A What is variation?

KEY FACTS

1 Variation is the name given to the differences between living things – most often when considering different individuals of the same species. For example, think about differences such as eye, hair and skin colour between people you know.

2 Some variation is genetic variation – it is caused by genes which are inherited by offspring from their parents.

- Most genetic variation between offspring happens because they inherit different combinations of their parents' genes [see meiosis on page 59].
- Some genetic variation happens because of a sudden change (mutation) in a sex-cell gene of the parent. Most new mutations are lethal (lead to death).

How are identical and non-identical twins formed?

Identical twins = same genetic material

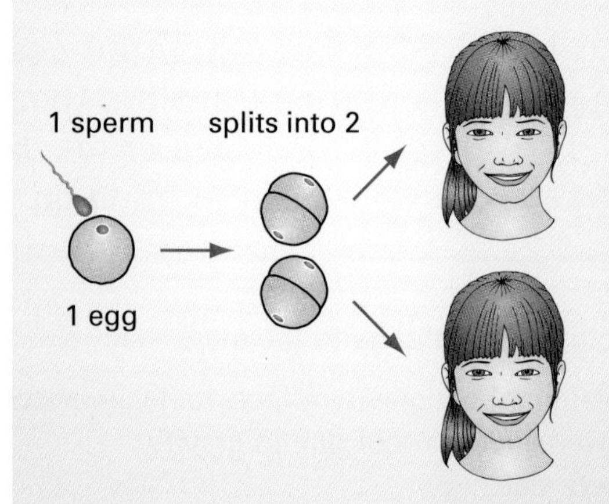

Brothers and sisters including non-identical twins grow from different fertilised eggs = different genetic material

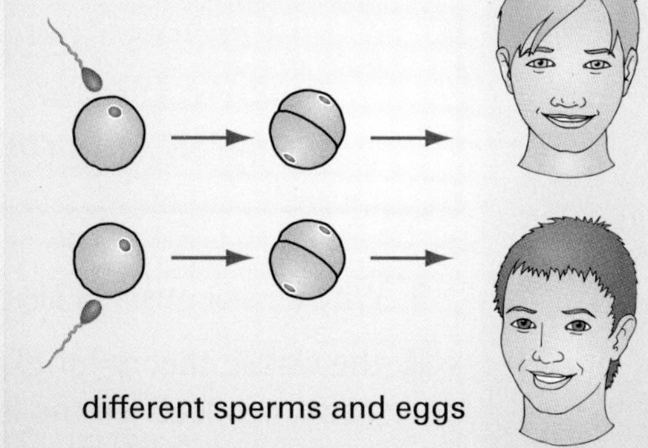

Q Dalmatian puppies may be very similar in appearance but they do not all have spots in the same pattern. How is this variation caused?

KEY FACT

3 Variation can also be due to environmental factors: nutrition, light, disease, and lifestyle factors such as exercise.

B About genetic material

1 Variation has been known about for a long time, but the **nature of the genetic material that controls** it was only discovered during the 20th century, when it was shown to consist of **DNA**.

KEY FACTS

2 DNA molecules make up most of the nucleus of every cell. The pattern of chemical units making up DNA acts as a code or set of instructions for cell processes.

3 DNA is organised into chromosomes.

Q What is the total number of chromosomes in
a) a sperm cell
b) a kidney cell?

Chromosomes are:

- always in **pairs in body cells** e.g. a muscle or skin cell. Each human cell has 23 pairs of chromosomes in the nucleus.
- always **single in sex cells** (called gametes) within sex organs. Human sex cells (sperm and ovum) have 23 single chromosomes.

KEY FACT

4 A gene is a short chunk of DNA which codes for a particular protein.

- Proteins give rise to particular characteristics. For example, the protein insulin provides the ability to control the level of sugar in the blood. However, in many cases, a characteristic is caused by a combination of more than one gene.

C Genomes

A **complete set** of all the genetic material in a living thing is called its **genome**.

Case study: The Human Genome Project

A huge project has been carried out across the world, to map the position and function of every human gene. This will help develop gene therapies for inherited diseases. For example, an aerosol spray can be used to carry genes for normal mucus production into the lungs of someone who has cystic fibrosis.

In theory, it is possible to screen a person's DNA to see if they carry problem genes that might cause difficulties for their offspring. A code of **ethics** (stating the rights and wrongs) must be put in place before this happens routinely. We need to think about the advantages and disadvantages of this type of screening.

Q Can you suggest one advantage and a disadvantage of screening for defective genes?

1 How does variation happen?

2 How can variation lead to both extinction and evolution?

Evolution and genetic engineering

THE BARE BONES

- ➤ The theory of evolution by natural selection gives an explanation of how new species can evolve.
- ➤ The theory of evolution developed over a period of years as more evidence became available, and people became used to new ideas.

A Evolution

A few hundred years ago, it was commonly believed that the Earth and living things were spontaneously created by God, as described in the Bible. This **creationist theory** implies that all life was **present at the start** of the Earth and that life forms have **not changed** since the beginning.

In 1809, Frenchman Jean Baptiste de Lamarck proposed that life evolves because of **changes** that happen to animals **during their life**. They **acquire** characteristics and can then **pass them on** (e.g. people who do physical work develop strong muscles, and so their offspring would have strong muscles too.) Lamarck's ideas were not accepted because people believed in creation, but he made people think about evolution. His ideas are not accepted today, because modern biologists follow the ideas of Darwin and Wallace.

Charles Darwin and Alfred Wallace, British naturalists, each developed a **theory of evolution by natural selection** in the late 1800s. They both based their ideas on years of studying living and fossil organisms. The main points of their theories are that:

- mutations cause **genetic variation** and give rise to **new characteristics**
- a new characteristic may be **beneficial or harmful** to the survival of an individual
- a **beneficial** characteristic is **more likely to be passed on** to future generations
- in time, this process may lead to **new species** evolving.

Darwin and Wallace wrote their theory before **genes** were understood. Their ideas were only fully accepted after their deaths, as peoples' views changed and scientists provided more evidence.

Q Can you distinguish between the theories of creation, acquired characteristics and evolution by natural selection?

Case study: The time scale of evolution

The main events are represented here as though they have taken place over a single year. Notice that modern humans appear very late in the timescale.

Jan 1	May	Aug	Nov	early Dec	mid-Dec	Dec 21	Dec 26	Dec 30	Dec 31 (late a.m.)
Earth formed	most land masses are present	bacteria occur	jawed fish arise; plants grow on land	reptiles evolve	birds evolve	egg-laying mammals appear	last living dinosaurs; primates evolve	apes first appear	modern humans evolve

C Genetic engineering

KEY FACT

1 Genetic engineering involves transferring genes artificially from one organism to another.

Example: There is a human gene which codes for insulin production.

- The gene is 'cut out' using enzymes, and transferred to a bacterium.
- Bacteria are cultured and produce insulin on a large scale.

Insulin is pure, safe for use in humans and available cheaply for diabetics.

2 There is some concern that genetically-engineered organisms may escape into the natural environment, and we do not know what the effects might be.

Q What are the advantages of genetically-engineering medicines?

D Case Study: Genetically-engineered growth hormone

Some people have a pituitary gland which does not produce sufficient growth hormone for the normal rate of growth, and so they don't grow to their expected height. This condition can be treated by injecting growth hormone at the correct phase of life.

Previously, the only source of growth hormone was from the pituitary glands of people who have died. Sadly, some people treated this way have contracted Creutzfeld-Jakob disease (CJD), because the hormone they were given was contaminated.

Nowadays, human growth hormone is produced by genetic engineering. This means there is a plentiful and safe supply.

Q What type of medicine is used to correct abnormally-slow growth rate?

PRACTICE

1 A particular type of moth exists in dark and light forms. In an experiment some years ago, moths were released in different areas of the UK and recaptured later. The table shows data.

area of UK	experiment	number of light-coloured moths	number of dark-coloured moths
unpolluted with soot	released	488	485
	recaptured	59	30
polluted with soot	released	62	157
	recaptured	5	83

a) What percentage of dark-coloured moths were recaptured in the two areas?

b) What main factor might affect the survival rate of dark-coloured moths in each area?

c) Predict the effect of reducing pollution on both types of moth.

2 How might genetic engineering be used to help make a gene therapy for treating people with an inherited disease?

How genes are passed on

- Living things may reproduce either sexually or asexually.
- Sexual reproduction results in more genetic variation.
- The main ways of influencing inherited characteristics are cloning, selective breeding and genetic engineering.

A Asexual reproduction

KEY FACTS

1. Asexual reproduction involves only one parent. The process involves an increase in cell numbers, and all the new cells are identical genetically (no variation).

2. Mitosis is the type of cell division that happens when many-celled living things are growing, and also when they are reproducing asexually.

Examples of asexual reproduction

Identical offspring are called clones.

Case study: Cloning

Plant cloning is now common and important commercially. It is reliable because the numerous offspring have the same features as the parent plant.

One method is taking shoot cuttings and encouraging them to root and grow into adult plants, as is done with geraniums. Another method is taking a small lump of tissue, which contains cells that have not yet specialised, and growing them in a culture medium. Orchids, oil palms and cauliflowers are cloned in this way.

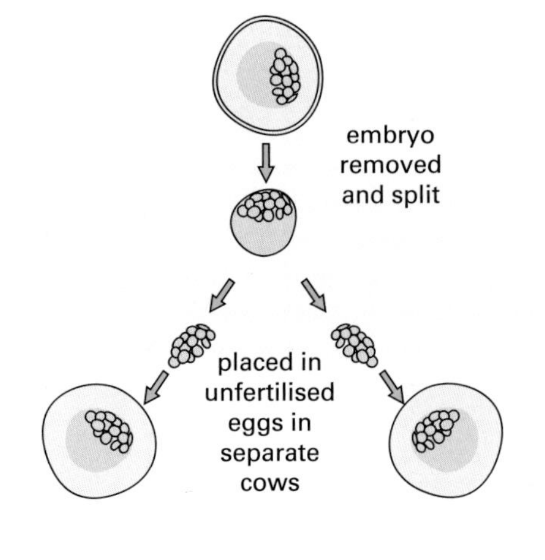

Cloning animals is not easy. It involves splitting the early embryo of a fertilised egg and putting each half into another egg cell. The egg cell is then placed in the womb of an adult female to develop.

Q A gardener takes a potato and cuts it into three pieces, each with a bud that will grow into a new plant. Explain why this is an example of asexual reproduction.

B Sexual reproduction

KEY FACT

1 Sexual reproduction involves fertilisation, in which the nucleus of one sex cell fuses with the nucleus of another.

2 Features of sexual reproduction:

- The sex cells produced by a parent are not all identical – there is **genetic variation** between them.
- Each offspring therefore inherits a **combination of genes** from its parents that is **different from other offspring.**
- Offspring generally **resemble their parents**, because they inherit their characteristics from them.

Q Why is:
a) an egg cell haploid?
b) a baby's skin cell diploid?

KEY FACT

3 Sex cells are produced when cells divide by meiosis.

- The number of chromosomes is **halved** (the sex cell nucleus is **haploid**).
- 'Swapping over' between chromosomes and parts of chromosomes during meiosis means that sex cells are not all genetically identical

4 Fertilisation restores the number of chromosomes to a **diploid** (double) **set**.

C Selective breeding

KEY FACT

1 Selective breeding means choosing only plants and animals with desirable characteristics to breed from.

- It is a **long-term** project because it can take many generations to breed offspring with the right combination of characteristics.
- It's **unpredictable**, because the exact combination of inherited characteristics is not know until the offspring is born.

2 **Selective breeding** is very important in agriculture because it has resulted in **increased yields** from crops plants and farm animals.

Breed 1: Large litter, but low survival rate for piglets

Breed 2: Small litter, but excellent survival rate for piglets

Breed 3: Large litter and high survival rate for piglets

Q Why has selective breeding been used by dog breeders?

PRACTICE

1 Say whether these are examples of asexual or sexual reproduction:
a) You dig up a clump of lilies, divide it into smaller clumps and replant several.
b) Frog spawn appears in the garden pond each Spring.
c) A vet fertilises a cow using sperm from the farmer's prize bull.

2 Suggest characteristics which farmers might find useful in selective breeding of:
a) cattle for milk production b) wheat for fast cropping.

Patterns of inheritance

THE BARE BONES

- Individual genes are inherited separately from one another.
- Simple patterns of inheritance can be worked out using diagrams.
- The basic laws of inheritance were worked out by Gregor Mendel in the 19th century.

A Basic principles

KEY FACT

1 Individual genes are inherited separately from one another, and are responsible for the inheritance of specific characteristics.

2 Some features, such as blue versus brown eyes, are controlled by a single gene, while others, such as height, are controlled by many genes.

3 Where only one or a few genes are involved, it is possible to predict patterns of inheritance using genetic diagrams.

Q How is it possible to predict patterns of inheritance?

B Gregor Mendel: founder of genetics

Gregor Mendel was a German monk and science teacher who, in the 1860s, discovered the basics of how inheritance happens. He carried out thousands of crosses between pea plants, studying seven different characteristics, including flower colour, stem length and colour of seed coat.

- Mendel started with plants that had only produced red or white flowers for several generations, so only contained one type of factor or allele.
- He crossed red and white flowered plants, took seeds and grew the offspring and then crossed these plants again, for several generations.
- All flowers were red or white (no pink flowers occurred).
- White flower colour disappeared in the first generation, but reappeared in the next.

Q Which piece of evidence made Mendel think that factors (or alleles) do not mix, but are inherited separately?

C Predicting inheritance

KEY FACTS

1 Some genes control a single, obvious characteristic, such as eye colour. Alleles are different forms of the same gene – e.g. an allele for blue eye colour or an allele for brown eye colour.

2 An offspring inherits two alleles for a characteristic, one from each parent.

C

Remember
An allele is one form of a gene, and two alleles are inherited for each characteristic.

- A **dominant allele**, even if there is only one of them, will **hide** the appearance of a **recessive allele**, – e.g. brown eye colour is dominant over blue eye colour
- Geneticists write a **dominant** allele with a **capital letter** and a **recessive** allele with a **lower case letter** – e.g. B stands for the brown eye colour allele and b stands for the blue eye colour allele.
- Someone is **homozygous** for a particular characteristic if they have inherited two alleles the **same** e.g. BB or bb.
- Someone is **heterozygous** for a particular characteristic if they have inherited two **different alleles** e.g. Bb.

KEY FACT

3 The genotype describes the types of alleles someone has for a characteristic, either heterozygous or homozygous. The phenotype describes how a gene shows itself – e.g. having blue eye colour.

Q What is the difference between the terms 'heterozygous' and homozygous'?

Example: Inheriting eye colour

Parents	mother	offspring	father
alleles for eye colour	bb		BB
colour of eyes	**blue**		**brown**
possible allele in sex cell	b		B
pair of alleles in offspring	Bb		

Eye colour of this offspring is brown, because brown eye colour is dominant over blue eye colour.

1 In Mendel's experiments, which piece of evidence suggests that one allele can be dominant over another?

2 Using Mendel's experiments, describe what happens when two red plants with mixed alleles (Rr) are crossed. What is the ratio of red and white flowered plants in the offspring?

3 In humans, some people have different ear lobe shapes.

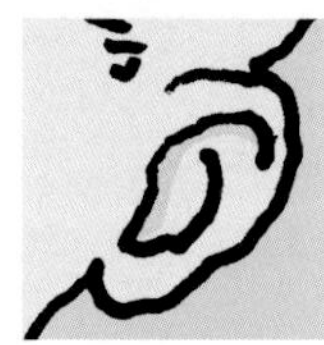

ear lobes droop (called 'free')

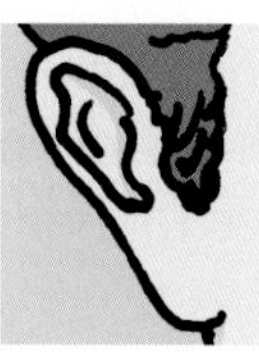

ear lobes slope (called 'attached')

What are the chances of two people having children with attached ear lobes, if their alleles are dd and Dd?

More about inheritance

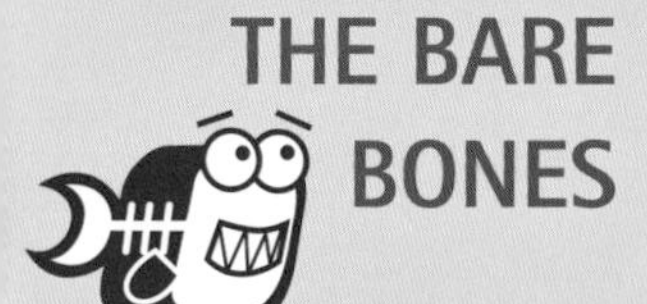

THE BARE BONES

- Gender is controlled by one pair of chromosomes.
- Diseases may be inherited.
- Punnet squares and family tree diagrams can be used to predict inheritance.

A Which sex?

KEY FACT

In humans, one pair of chromosomes controls gender, XX in females and XY in males. These are the sex chromosomes.

In humans, males can be distinguished from females by looking at the sex chromosomes. The Y chromosome is shorter than the X chromosome it is paired with.

Humans have 22 pairs of ordinary chromosomes and 1 pair of sex chromosomes. Some genes only occur on the sex chromosomes and so they are said to be sex-linked.

	female	male
alleles in parent	XX	XY

female XX → X, X; male XY → X, Y

Offspring: XX, XX, XY, XY

Offspring are 50% male and 50% female.

Q Suggest why the 50:50% ratio of males and females is advantageous.

B Disease and family trees

Haemophilia is an inherited disorder in humans, in which a blood-clotting factor is not produced. The allele for normal clotting (H) is dominant to (h), the allele for abnormal clotting. The family tree (below) shows the pattern of inheritance for haemophilia.

female
male
has haemophilia

1 2
3 4 5 6 7 8
9 10 11 12 13 14 15
16 17 18 19 20

Q What is the percentage of haemophiliacs in the tree opposite?

C Inheriting diseases

KEY FACT

Inherited diseases tend to be recessive. Because of this they can be carried by members of a population without symptoms showing, as the normal gene is dominant.

1 Cystic fibrosis and haemophilia are two examples of recessive inherited diseases.

2 Huntingdon's disease is a disorder of the nervous system that can be passed on by just one parent because it is caused by a dominant allele.

3 Sickle-cell anaemia is caused by a recessive allele, which results in abnormally-shaped haemoglobin and distorted red blood cells. However, carrying one allele for sickle cell can help protect people against malaria.

Q How likely is it that two people will have a child with cystic fibrosis, if their alleles are Cc and CC (where c = cystic fibrosis and C = normal)?

Case study: Cystic fibrosis

Around 50% of Europeans carry one allele for cystic fibrosis, but suffer no effects as it is a recessive allele. Someone with cystic fibrosis produces sticky mucus in the lungs and gut. The lungs get blocked up easily and fluid collects, and there are problems with digestion.

Which examples of genetic diseases do you need to know for your exam board?

PRACTICE

1 Looking at the family tree in section B:

a) how many alleles for abnormal clotting must be inherited for haemophilia to occur in females?
b) how do you know that females in the family tree must be carriers even though they do not have haemophilia?
c) what percentage of the family members shown had haemophilia?
d) suggest why one treatment for haemeophilia involves injecting clotting factor?

2 The family tree below shows the inheritance of freckles for a particular family.

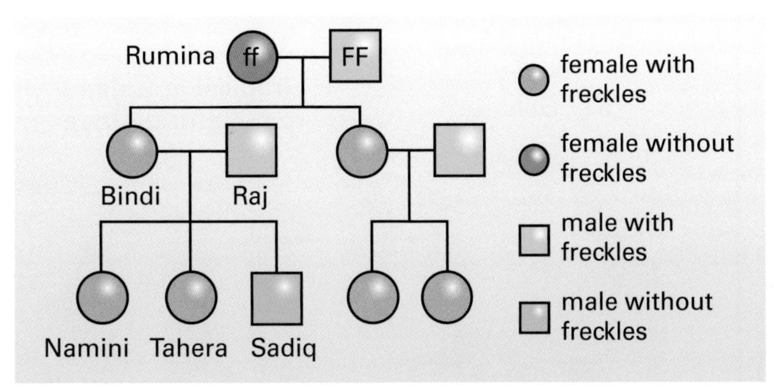

a) How can you tell that Rumina is homozygous for this characteristic?
b) How can you tell from his children that Tariq probably has two alleles FF for freckles? What other combination of alleles might he have?
c) What combination of alleles did Sadiq inherit for this characteristic?
d) Explain why Bindi and Raj must both have inherited the alleles Ff for freckles.

Ecosystems and biodiversity

- Major types of ecosystem on Earth include deserts, tundra and tropical rainforest.
- Resources in ecosystems are finite.
- Only careful management of the environment will safeguard the future of the planet.

A Ecosystems

KEY FACT

An ecosystem is made up of living things, the environment they live in and the interactions between them.

- The Earth is the largest ecosystem, but the climate and shape of the land vary around the world, giving several major terrestrial ecosystems:

Coniferous forest
e.g. Canada

- cold to mild
- acidic soil
- birds and mammals

Temperate forest
e.g. Northern Europe

- mild (no extremes)
- large trees, birds and mammals

Tropical rainforest
e.g. Brazil

- warm, very wet
- huge bio-diversity, rapid plant growth

Desert
e.g. Sahara

- hot daytime, cold night-time
- dry
- sparse vegetation

Tropical grassland
e.g. Central Africa

- hot and dry
- some regular rainfall
- grazing animals and their prey, invertebrates

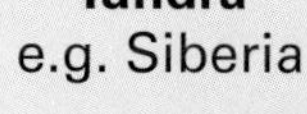

- extreme cold
- water locked as ice
- migrating birds/animals

Q Which major type of ecosystem, covering much of the Earth's surface, is not shown in the diagram on the right?

B Biodiversity

KEY FACT

1 Each ecosystem supports a different variety of living things. This is called biodiversity.

Biodiversity is **important** because:

- it represents the total pool of **genes**, as well as the potential for new ones.
- humans get many **resources** from living things, including **medicines**.

2 **Biodiversity** is **reduced** by the **environmental impact of many human activities.**

3 Within an ecosystem, living things **compete** for factors such as **food, space, minerals and light. Predation** and **disease** can also affect population size.

Case study: Rainforest biodiversity

In an area of the Malaysian rainforest, the biodiversity of birds was monitored between 1978 and 1980. In January 1978, some of the timber was logged and the forest was left to regenerate. The graph shows the biodiversity data for this time.

- The altitude of the land varies across the rainforest region – trees in one part tend to grow taller. This effect increases biodiversity in the area as a whole, because different species flourish at different heights in trees, and also different trees grew at different altitudes, so there was a greater variety of habitats.

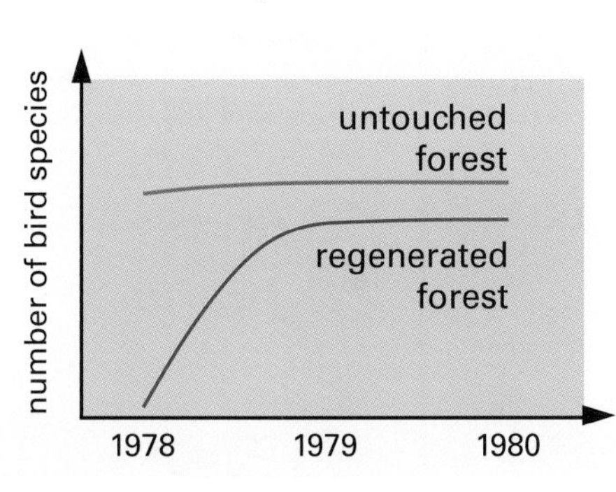

- Logging reduced the biodiversity and the number of bird species.
- By only logging trees with a trunk diameter of over 50 cm, regeneration occured more quickly as the younger trees were still there and able to grow.
- By comparing the data between the two areas, it can be seen that biodiversity did not completely return to normal within the time span.

Q Why is the rainforest seen as a vital resource for the world?

1 What are the main components of an ecosystem?

2 Suggest one reason why it is important to maintain biodiversity.

3 a) List three factors that the crop plants are competing for.

b) Suggest an environmental factor that might explain the growth pattern seen in this field of wheat.

c) How might a farmer influence the growth pattern of this crop?

Human impact on environment

- ➤ Waste products can cause long-term pollution.
- ➤ There are competing priorities for using and conserving the Earth's resources, which have to be balanced by considering the benefits of using them and the environmental damage that can occur.

A Human activities have an impact

1 When the human population was small, its impact on the environment was small and local. Today, however:

- humans **exploit resources** faster than other species, especially **non-renewable** resources, which cannot be replaced.
- the human **population is growing** very fast and the way we live has changed.

Agriculture – through intensive farming, requiring the use of agrochemicals, e.g. inorganic fertilisers and pesticides and recently, genetic engineering.

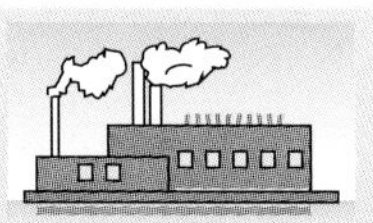

Industry – getting the raw materials, processing them and distributing products all generate pollution.

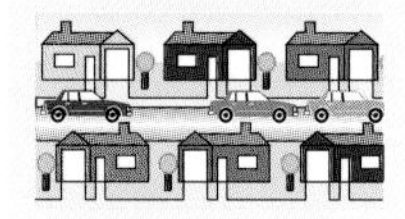

The growth of cities – requiring space, and generating pollution due to increased traffic and domestic heating.

Drastic changes to the ecosystem – such as deforestation, draining marshland, mining, damming and flooding natural areas, or redirecting waterways.

2 **Agriculture** is a kind of artificial ecosystem. Farmers may maximise food production by using:

- fertilisers
- pesticides
- antibiotics in foodstuff

KEY FACT

3 Many human activities cause pollution, which is one reason why they have an undesirable impact on the environment.

Long-term pollution, such as heavy metals in groundwater supplies, or radiation, can have unpredictable effects and are not easy or always possible to reverse.

Types of pollution

Fossil fuels	burning fossil fuels: oil, gas and coal for heating and transport, produces: • carbon dioxide, which is a greenhouse gas and causes global warming • sulphur and nitrogen oxides, which cause acid rain
Chemicals	manufacturing processes make toxic wastes modern farming uses pesticides and fertilisers: • polluting waterways and air
Sewage	untreated sewage may be fed into the sea: • carrying diseases

Q What is the main reason for the increase of human impact on the environment?

B Protecting the planet

Many of the Earth's resources are **non-renewable**. Choices must be made about how to use them, but there are often **competing priorities**.

Case study: Fishing for food?

One of the Earth's richest resources is the sea. In the past, people successfully fished in the North Sea and around the Grand Banks off the east coast of Canada. In recent years, improved fishing methods meant bigger catches, especially of cod. Despite controls, fish stocks dropped to very low levels and now there is a ban on cod fishing. Scientists hope this will give remaining cod time to grow and reproduce, replenishing stocks. Alternative strategies might include:

- fish farming e.g. of salmon and trout
- growing protein-rich foods, e.g. pulses and beans, as alternative food sources.

Q Why might each of the following people be interested in how fishing in the North Sea is controlled:
a) a fisherman?
b) a shopper?
c) a scientist?
d) a politician?
e) a farmer?

There are many strategies for protecting the future of our planet:

Recycling materials such as glass, aluminium and paper helps reduce waste and the need for new materials.

Special programmes aimed at **protecting endangered species** which face extinction e.g. CITES (Convention on International Trade in Endangered Species); captive breeding; national parks and seed banks.

Managing ecosystems e.g. felling only a small proportion of trees in a forest on a rotation basis, rather than a whole forest area; growing specialist crop plants in small areas within a rainforest, rather than clearing the whole area for farming.

Conservation programmes help protect **sensitive ecosystems** e.g. Sites of Special Scientific Interest.

Improve public transport so that the use of private vehicles can be reduced.

Use purchasing power to encourage companies to produce goods with less packaging.

Increase education aimed at influencing the attitudes of future generations.

Use international pressure in the form of subsidies and penalties to encourage governments to manage local environmental issues properly.

Local recycling schemes encourage the separation of waste by each household, so that more materials can be reused.

PRACTICE

1. Why does a change in the size of human population affect the environment?
2. What are non-renewable resources?
3. Explain what is meant by 'competing priorities' by using an example from the case study above.
4. Why does burning fossil fuels cause pollution?

Survival and interdependence

THE BARE BONES

- Adaptation to the environment is necessary for plants and animals to survive.
- Despite competition between living things, there is also an interdependence.

A Adaptations and survival

KEY FACTS

1 Survival involves getting the basic needs of life, including food, protection from natural enemies, and shelter from unfavourable conditions, such as a harsh climate.

2 Living things are adapted differently to survive in a range of ecosystems, and this gives them an advantage over others that are not.

Adaptation	Arctic fox	desert rat
body size and surface area	large body size, but smaller surface area compared to body size = less heat loss	small body size, but larger surface area compared to body size = more heat loss
thickness of insulating coat	thick coat	thin coat
amount of body fat	thick layer	thin layer
camouflage	white coat	brown coat

Q The desert rat produces little or no urine. Suggest how this helps it survive in its environment.

3 Survival rate is influenced by adaptation to the environment, as well as by interdependence and competition with other living things.

B Competition and interdependence

KEY FACT

1 All the basic needs of life, such as food and minerals, available light and space, come from the environment. This means that resources are finite – there is only so much of any resource. Living things compete for these resources.

B

- Some resources, such as minerals, are **naturally recycled** when living things die.
- Light is a **non-renewable** energy source, but unlikely to 'run out' in the imaginable future.

2 Living things are **linked through their need for food. Food chains** describe the feeding relationships between them [see *also page 72*].

- **Photosynthetic** organisms – mainly plants – are always **at the start** of a food chain because they are food **producers**.
- **Animals** are **consumers**, because they eat ready-made food in the form of plants or other animals.
- Many food chains overlap, forming a **food web**.

Example: wheat crop → rabbit → fox → invertebrates → microbes

Remember
In a food chain the arrows always lead <u>from the food to the feeder</u>.

KEY FACT

3 A change to the population size of one species will influence the populations of other living things that feed on them.

Q List the producers and consumers shown in the image on the right.

It is common for questions to ask what happens if one factor in an ecosystem changes. Think about the living things that are immediately affected e.g. if the number of foxes increases, the number of rabbits decreases. But also think about further links e.g. fewer rabbits may mean less crop damage.

1 What does adaptation to environment mean?

2 Why is the ratio of surface area: body volume important for small mammals?

3 Grasses that are adapted to living in very dry conditions tend to have few stomata. How does this help their survival?

4 Imagine that pollution kills most of the water beetle population. What would happen to the other organisms in the food chains that include water beetles? Explain the 'knock-on' effects.

Predators, prey and resources

THE BARE BONES

- The population sizes of predator and prey are linked.
- Nature recycles resources such as carbon and nitrogen.
- Biological control uses a natural predator to reduce pest populations.
- Humans affect the carbon cycle by increasing the level of CO_2 in air.

A Predation

KEY FACT

A predator is an animal that eats another, and the prey is the animal which is eaten. The sizes of the two populations are interdependent.

Case study: Predator and prey

The populations in the image below show a typical shape as each population rises and falls, the shape of the predator graph following the shape of the prey graph. There may also be a seasonal effect. For example, a hard winter may reduce the number of ladybirds the following spring, so the greenfly population increases faster than the year before. Biological control uses a natural predator to control the population of an organism that is a pest to a farmer – e.g. parasitic wasps to control aphids. The pest is kept at low levels which cause little commercial loss, but is not eradicated.

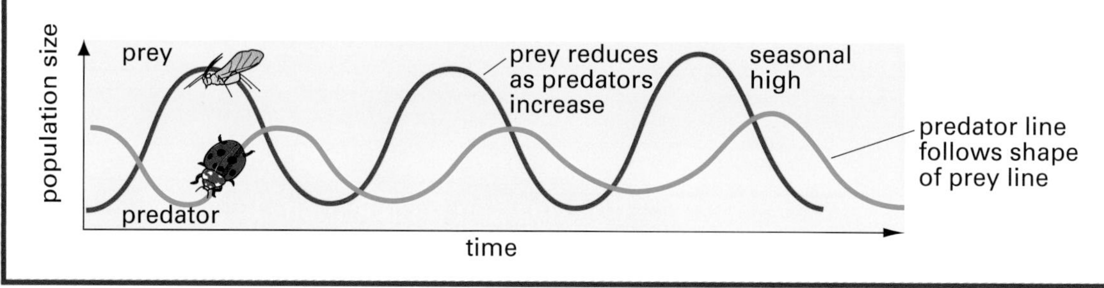

Q What is biological control?

B Natural cycles

The use of an ecosystem's resources can often be described in terms of cycles. Two of the most important are the carbon cycle and the nitrogen cycle.

KEY FACT

Nitrogen is necessary for building cells of living things, and a lack of it can limit growth.
Nitrogen is present in the air as insoluble nitrogen gas, in soil water as soluble nitrate, and in protein in living things.

The carbon cycle is shown on the opposite page. Note how carbon is recycled in the different processes and learn them.

B

1 **Carbon** makes up a large proportion of living cells. It is present in:

- **biomass** (the bodies of living things) and their remains, including **fossil fuels**
- the air as **carbon dioxide**
- the rocks which contain carbonate

2 Carbon dioxide is added to the atmosphere by living things when:

- they carry out **respiration**
- things **decay**
- we **burn fossil fuels** or other materials from living things, e.g. wood

3 Carbon dioxide is removed from the atmosphere by plants during **photosynthesis**.

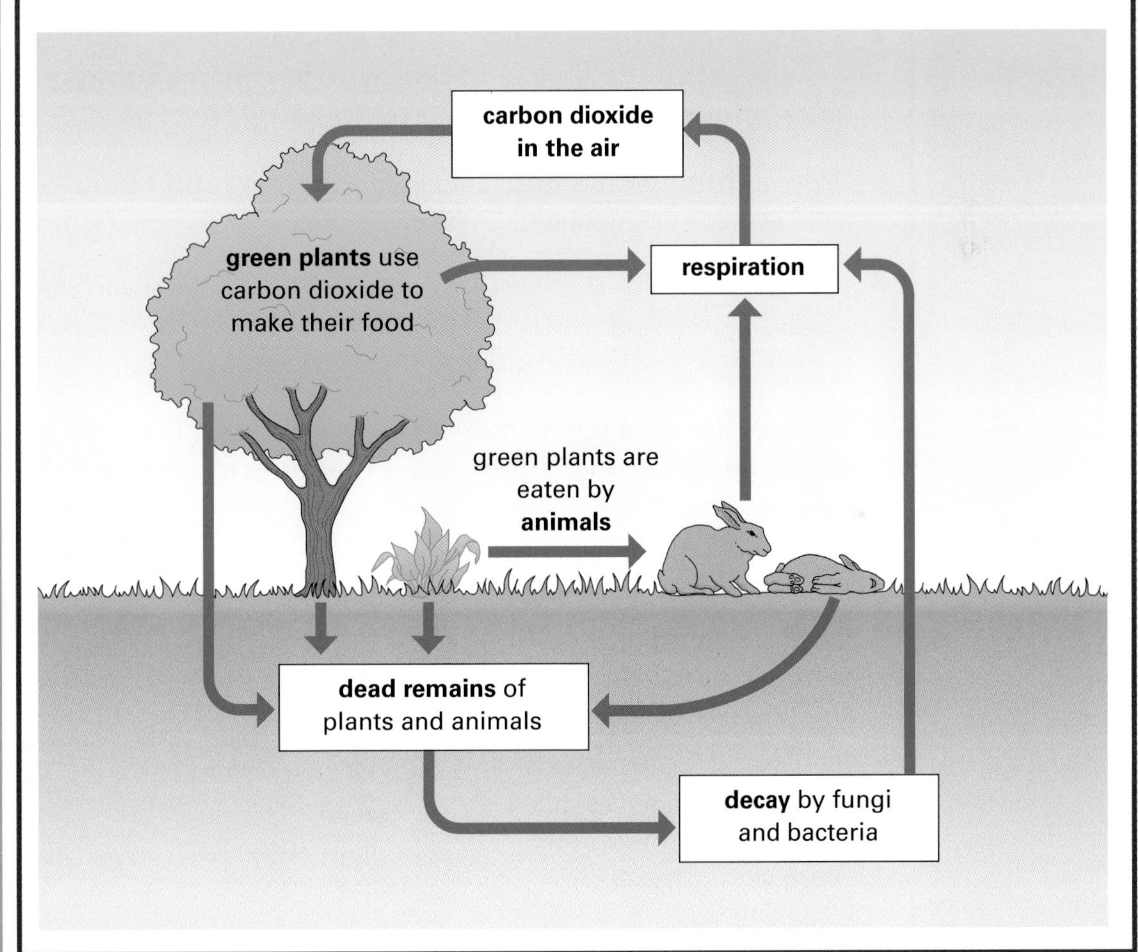

Q Why are these natural cycles important to the environment?

1 How are populations of predator and prey interdependent?

2 Large cane toads were introduced into Australia from South America, to control rats in sugar cane plantations. The toads are big killers and eat a variety of other prey. Suggest undesirable effects on the ecosystem of introducing this predator.

3 How does burning fossil fuels affect the carbon cycle?

4 What are the roles of microbes in the following processes:
a) nitrogen fixing?
b) nitrifying?
c) denitrifying?

Food chains and energy flow

THE BARE BONES

- ➤ Food chains describe 'what eats what' in an ecosystem.
- ➤ Energy flows through the different levels of a food chain.
- ➤ A shorter food chain means less energy is wasted.

A Food chain basics

KEY FACT

1 The animals and plants in an ecosystem are interdependent – each either eats or is eaten by other living things. This pattern of interactions is called a food chain.

- The term **food web** is sometimes used as well as **food chain** – especially for a more detailed description of a real ecosystem.

KEY FACT

2 The stages of a food chain are called trophic levels. Plants (and bacteria which can photosynthesise) are producers and are always the first trophic level.

3 Producers use the **energy of the Sun** to make food. This is the **ultimate source of energy for living things**, and is **temporarily stored in biomass**.

KEY FACT

4 Animals are consumers as they eat plants and other animals.

- The **second trophic level** is made up of animals called **primary consumers**, which are **herbivores** (eat plants).
- The **third trophic level** is made up of animals called **secondary consumers**, which are **carnivores** (eat animals).

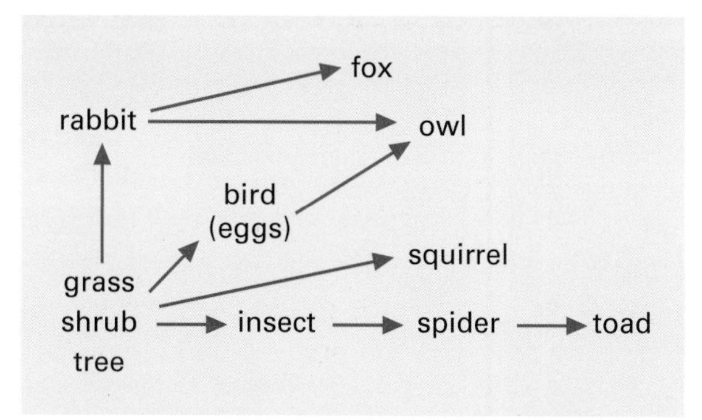

5 **Decomposers** – living things that get their food by breaking down **dead plant and animal material** – are also important in many food chains.

Q What is the maximum number of trophic levels shown in the food web (left)?

B Pyramids of biomass

The mass of living material at **each stage in a food chain** can be **drawn to scale** and shown as a **pyramid of biomass**:

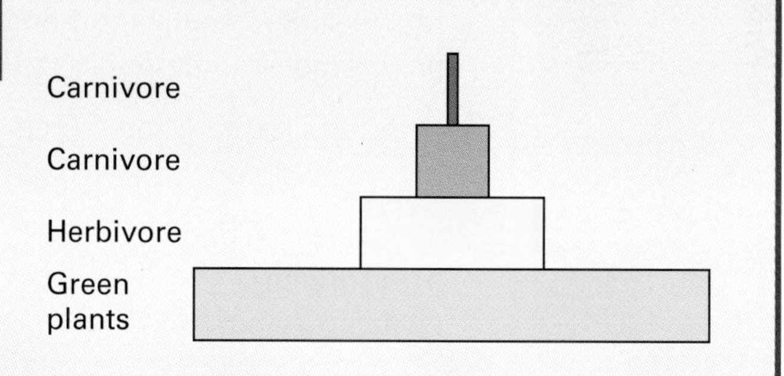

B

Example: Drawing a pyramid of biomass

In the data below, the producers will include oak trees, brambles, grass, bluebells, other shrubs and smaller plants. Primary consumers include worms, rabbits, field mice, etc. Other consumers include stoat, birds, fox, badgers, etc.

Data for an English woodland

Trophic level	g/m^2
1 (producers)	6000
2 (primary consumers)	3
3	1.75

1 decide on a scale e.g. 1 kg = 1 cm on the bar representing the trophic level
2 convert to data in kg to cm e.g. 500 g = 1 cm so 6000 g = 6000 ÷ 500 = 12 cm
3 draw to scale, with producers at the base

Q Why does the size of each trophic level decrease in the food chain?

C Food production and energy flow

KEY FACT

1 At each stage in a food web, less material and less energy are available to be stored in biomass. Some is lost in waste materials, in repair of cells, or as heat during respiration.

The amount of energy that flows through a food chain depends on the plants and animals involved. In terms of food production, **shorter** food chains waste less energy.

Energy transfers through food chains

food chain	example	approximate energy yield to humans in kJ per thousand hectares
crop → human	wheat	10 000
intensive grassland → livestock → human	intensive cattle on grass	meat: 340 milk: 4 000
grassland and crops → livestock → human	mixed dairy farm	400

In terms of food production, the energy yield is greater from plant-based food chains, but the nutrient quality and digestibility is lower.

Q Why are short food chains more energy efficient?

1 Plankton are small plants and animals near the sea surface. Scientists found that the plants in each square metre used 100kJ of the energy they absorbed each day, in making new cells.
a) Suggest what happens to the rest of the energy.
b) The animal plankton gained 25 kJ of energy and used 10kJ to build biomass, and 12 kJ for moving. How much energy do they transfer as heat to the surroundings per day?

2 Draw a pyramid of energy using the same method as for a pyramid of biomass, using this energy data taken from a river in USA.

Trophic level $kJ/m^2/year$: 1 88 000 2 14 000 3 1 600 4 100

Atomic structure

- ➤ Atoms are made of subatomic particles: protons, neutrons, and electrons.
- ➤ Atoms can be defined in terms of atomic number and mass number.

A Basics of atomic structure

Q State the names and location of the sub-atomic particles.

1. Atoms have a heavy nucleus made up of protons and neutrons.
2. Electrons circulate around the nucleus in elliptical orbits.
3. Protons, neutrons and electrons are subatomic particles. These have electrical charge and mass.

B Mass and charge

Remember
Protons are plus, neutrons are neutral, and electrons are negative.

1. Protons and neutrons both have a mass of 1.

- This mass is called one atomic mass unit (1u) and is equal to 1.67×10^{-24}g.

2. Electrons have a mass of $\frac{1}{1840}$ of an atomic mass unit (0.00055u). This is negligible and generally ignored.
3. Protons each carry a single positive charge and electrons each carry a single negative charge.
4. Neutrons carry no charge and are neutral.
5. To sum up:

KEY FACT

Particle	Symbol	Mass	Charge
proton	p	1	+1
neutron	n	1	0
electron	e	0.00055 (negligible)	-1

6. The attraction between positive and negative charges holds the electron in place around the nucleus.

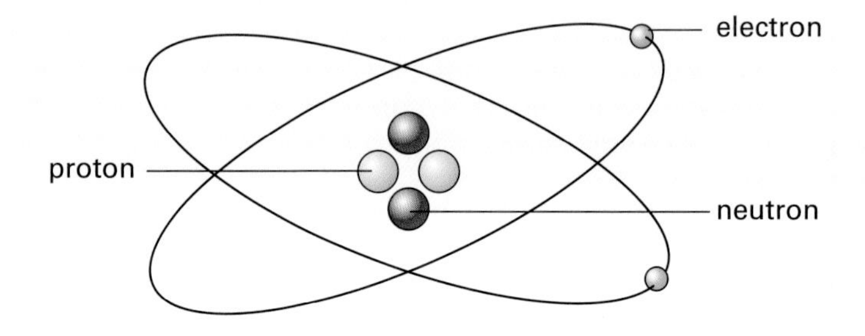

Q What is meant by saying that atoms are 'neutral'?

C Atomic number and mass number

1 All atoms are **electrically neutral** because:

KEY FACT

the number of protons = the number of electrons

2 The number of protons (and electrons) in an atom is the **atomic number (Z)** of that atom.

- The atomic number (or proton number) is **unique to an element** and identifies it.
- **Example:** only magnesium atoms contain 12 protons and thus only magnesium has an atomic number of 12.

3 The total number of protons and neutrons in an atom is its mass number (A), hence

KEY FACT

number of protons + number of neutrons = mass number

4 **Example:**

- Sodium atoms have 11 protons and 12 neutrons in the nucleus.
- Hence the atomic number (Z) of sodium (Na) is 11 and the mass number (A) is 23.
- The atom can be written as:

mass number ⟶ $^{23}_{11}Na$ ⟵ symbol
atomic number ⟶

5 Similarly, chlorine atoms have 17 protons and 18 neutrons in the nucleus.

- The atomic number of chlorine is thus 17; the mass number is 35.
- The atom is $^{35}_{17}Cl$.

Q What is the mass number a measure of?

PRACTICE

1 Which sub-atomic particle:
a) is neutral
b) has negligible mass
c) is negatively-charged
d) has the same mass as a neutron?

2 Explain why the circulating electrons do not fly away from the nucleus.

3 How many protons, neutrons and electrons are there in one atom of:

a) $^{16}_{8}O$

b) $^{27}_{13}Al$

c) $^{235}_{92}U$

d) $^{35}_{17}Cl$

e) $^{64}_{29}Cu$

4 Explain the difference between 'atomic number ' and 'mass number'.

Isotopes and mass

THE BARE BONES

- All the atoms of any one element are not identical to each other because they may contain different numbers of neutrons.
- Atoms of the same element but with different numbers of neutrons are called isotopes.
- The 'relative atomic mass' quoted for an element takes into account its different isotopes.

A Isotopes

Remember
Isotopes are different atoms, but of the same element.

1 All the atoms of an element have the same number of protons and hence the same atomic number.

2 Because they have the same number of protons they must have the same number of electrons.

3 As the chemical reactions of an atom depend upon the number of electrons it contains (see later topics on bonding), then all the atoms of an element have the same chemical properties.

4 However, atoms of an element can have different numbers of neutrons and hence different mass numbers. These are isotopes.

KEY FACT

5 Isotopes are atoms of the same element with the same number of protons (atomic number) but different numbers of neutrons and hence different mass numbers.

Q State the similarities and differences between the isotopes of an element.

6 **Example:** The two isotopes of carbon are:

$^{12}_{6}C$ i.e.

6 electrons

6 protons, 6 neutrons (mass number 12)

$^{14}_{6}C$ i.e.

6 electrons

6 protons, 8 neutrons (mass number 14)

KEY FACT

7 Isotopes have the same:
- numbers of protons
- numbers of electrons
- atomic number
- chemical properties

Hence the proportions of the isotopes in a sample of the element do not affect its chemical reactions.

KEY FACT

8 Isotopes have different:
- numbers of neutrons
- mass numbers
- physical properties, such as density, melting/boiling point

As you will see in section B, the proportions of the isotopes in a sample of the element do affect its physical characteristics.

B Relative atomic mass

1 Because both protons and neutrons weigh 1u, the mass number of any single atom must be a whole number.

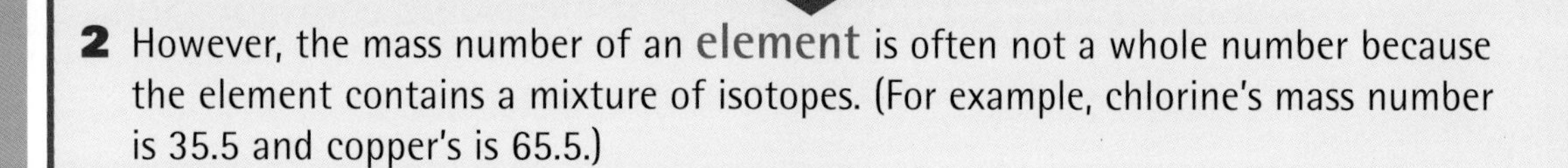

2 However, the mass number of an **element** is often not a whole number because the element contains a mixture of isotopes. (For example, chlorine's mass number is 35.5 and copper's is 65.5.)

3 This number is the **relative atomic mass** of the element and it takes into account the proportions of each isotope.

4 Example:

- Natural chlorine gas contains 75% of $^{35}_{17}Cl$ and 25% of $^{37}_{17}Cl$
- So, in every four atoms of chlorine, 3 have a mass of 35 and one a mass of 37.
- The average atomic mass is thus $\frac{(3 \times 35) + (1 \times 37)}{4}$ which is 35.5.
- The relative atomic mass of chlorine is therefore 35.5.

5 Similarly, physical properties, such as melting point and boiling point, are affected by the proportion of the isotopes.

- For example, 100% $^{35}_{17}Cl$ is lighter than 100% $^{37}_{17}Cl$ and would have a lower melting and boiling point.
- A 50:50 mixture has melting and boiling points mid-way between the two.
- In practice, the effect is very small and largely ignored.

Q Why is the relative atomic mass of some elements not a whole number?

PRACTICE

1 Oxygen has three isotopes: $^{16}_{8}O$, $^{17}_{8}O$, $^{18}_{8}O$.

State the similarities and differences between them.

2 Naturally-occurring neon contains 90% of the isotope $^{20}_{10}Ne$ and 10% of the isotope $^{22}_{10}Ne$.

What is the relative atomic mass of naturally-occurring neon?

3 Given the following relative atomic masses:

H = 1, C = 12, O = 16, Ca = 40

What is the molecular mass (total mass) of each of the following molecules

a) H_2O b) CO c) $CaCO_3$? Show your working.

Electronic configuration

THE BARE BONES

- ➤ Electrons move in orbits round the nucleus of an atom.
- ➤ A set of orbits with similar energy is called a shell or energy level.
- ➤ Each element has a particular electronic configuration.

A Orbits and energy levels

KEY FACTS

1 Electrons in an atom are **moving constantly** in elliptical paths called **orbits**.

Electrons in orbits far from the nucleus have more energy than those in orbits nearer to the nucleus.

2 Each energy level (shell) has a maximum number of electrons it can contain.

- The **first** (lowest) energy level holds up to **2** electrons.
- The **second** holds up to **8** electrons in two sub-levels (holding 2 and 6 electrons).
- The **third** has **three sub-levels** (holding 2, 6 and 10 electrons - maximum 18).
- The **fourth** has **four sub-levels** (holding 2, 6, 10 and 14 electrons - maximum 32).

3 For convenience, energy levels are shown as **circles around the nucleus**:

4th shell : 32 electons
3rd shell : 18 electrons
2nd shell : 8 electrons
1st shell : 2 electrons

Q Explain the arrangement of electrons in the fourth energy level.

B The first eighteen elements

KEY FACT

1 Energy levels are filled in order, lowest energy level first.

2 The final arrangement of electrons in an atom is the **electronic configuration**.

3 Example:

- The aluminium atom $^{27}_{13}Al$ has 13 protons and therefore 13 orbiting electrons.
- There must be: 2 electrons in the first energy level, 8 in the second and 3 in the third.
- So the electronic configuration of aluminium is 2.8.3.

Remember
Fill energy levels from the inside, working outwards.

B

4 The electronic configurations of the first 18 elements are:

$_1$H 1	$_6$C 2.4	$_{11}$Na 2.8.1	$_{16}$S 2.8.6
$_2$He 2	$_7$N 2.5	$_{12}$Mg 2.8.2	$_{17}$Cl 2.8.7
$_3$Li 2.1	$_8$O 2.6	$_{13}$Al 2.8.3	$_{18}$Ar 2.8.8
$_4$Be 2.2	$_9$F 2.7	$_{14}$Si 2.8.4	
$_5$B 2.3	$_{10}$Ne 2.8	$_{15}$P 2.8.5	

Q What is meant by 'electronic configuration'?

C Elements 19-31

1 Element number 19 is $_{19}$K. It has 19 electrons in configuration 2.8.8.1

KEY FACT

2 The 19th electron goes in the fourth shell because the first two electrons of shell four have lower energy than the set of ten in the third sub-level of shell three.

3 The twentieth element is $_{20}$Ca with configuration 2.8.8.2

KEY FACT

4 The 21st element is $_{21}$Sc with configuration 2.8.9.2 because, after the first two of the fourth shell, the next lowest energy level is the third sub-level of shell three.

5 The process continues:

$_{22}$Ti 2.8.10.2	$_{25}$Mn 2.8.13.2	$_{28}$Ni 2.8.16.2
$_{23}$V 2.8.11.2	$_{26}$Fe 2.8.14.2	$_{29}$Cu 2.8.17.2
$_{24}$Cr 2.8.12.2	$_{27}$Co 2.8.15.2	$_{30}$Zn 2.8.18.2

(Elements Sc → Zn are **transition metals** - see page 112.)

Q Which are the next two energy levels to be filled after the first sub-level of shell three?

6 The third shell is now complete and the fourth shell is resumed at $_{31}$Ga with configuration 2.8.18.3

Know how to work out the electronic configurations up to 2.8.18.8 from the atomic numbers.

PRACTICE

1 Copy and complete the following table:

element	mass number	atomic number	number of protons	number of neutrons	number of electrons	electronic configuration
sodium	23		11			
magnesium		12		12		
iron				30	26	
beryllium	9				4	

2 Draw sketches to show the complete atomic structure of:

a) $^{1}_{1}$H b) $^{4}_{2}$He c) $^{40}_{20}$Ca d) $^{48}_{22}$Ti e) $^{70}_{31}$Ga

Ionic bonding

THE BARE BONES

- Sometimes atoms react together to form ionic bonds.
- In an ionic bond, electrons transfer between atoms.
- Ionic compounds form giant, three-dimensional structures (crystals).

A How an ionic bond is formed

1 Bonds between atoms are formed by electrons **transferring completely from one atom to another** (this Section) or by atoms sharing electrons (Section B).

2 The way that atoms react **depends upon their electronic configuration**.

- Only the **electrons in the outer shell** are involved.

KEY FACTS

Atoms lose or gain (or share - see Section B) electrons to attain a complete outer shell of 8 electrons.

3 An ionic bond (or electrovalent bond) is formed when electrons are lost and gained by the atoms involved.

4 All compounds containing metals are formed by ionic bonds.

Remember
Atoms are neutral because the number of positive protons equals the number of negative electrons.

- The metal atom loses its outermost electrons. It now has more protons than electrons and is positively charged (a positive **ion**).

5 The **non-metal atom gains** these electrons into its outermost shell and becomes a **negative ion**.

6 **Example:** forming sodium chloride from sodium and chlorine:

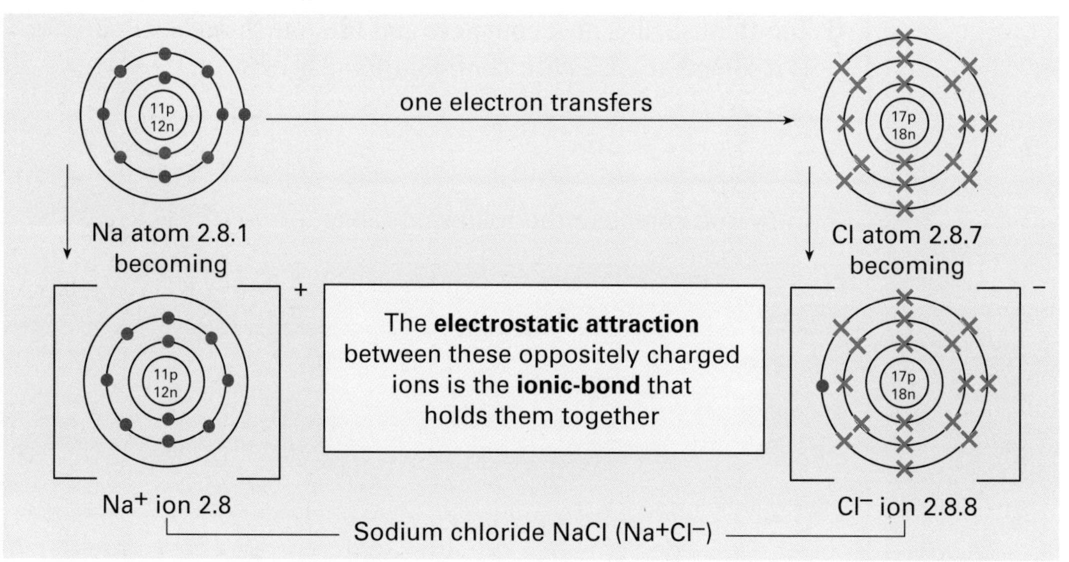

Q What is an ion?

KEY FACT

7 Note that the numbers of positive or negative charges on an ion are equal to the numbers of electrons lost or gained.

B Multi-charged ions

KEY FACT

Remember
All electrons are alike, even though, for clarity, they are represented differently in these 'dot and cross' diagrams.

Q Explain why a magnesium ion carries a 2^+ charge.

1 If more than one electron is lost/gained then the resulting ion will carry more than one positive/negative charge.

2 **Example:** the formation of magnesium oxide MgO (Mg^{2+} O^{2-}):

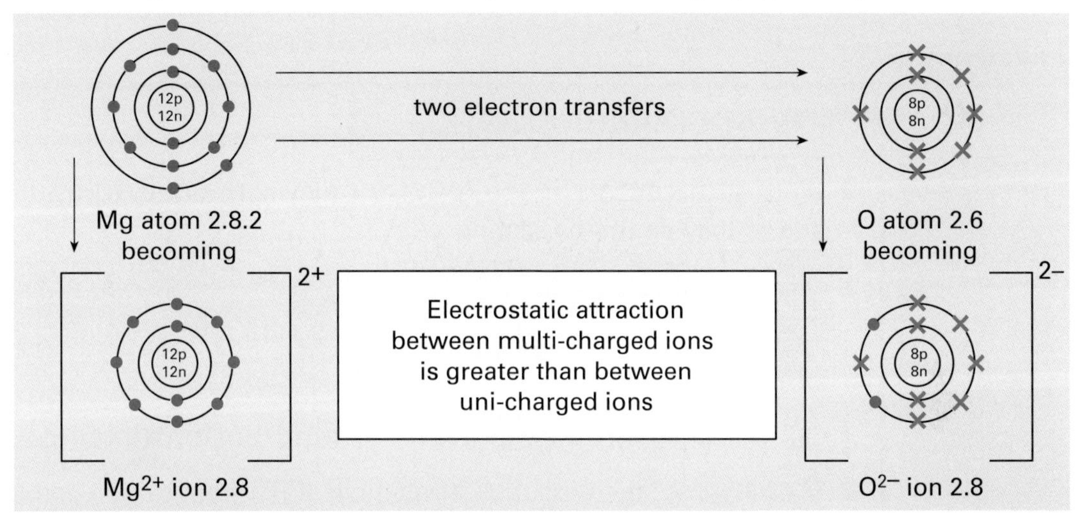

C Properties of ionic compounds

Remember
Any compound containing a metal will be an ionic compound.

Q What is meant by 'electrostatic attraction'?

1 In a sodium chloride crystal each Na^+/Cl^- ion is surrounded by 6 Cl^- / 6 Na^+ ions.

2 The result is a giant, three-dimensional structure held together by the electrostatic attraction between the ions.

3 The crystal is neutral because the total number of Na^+ and Cl^- ions is equal.

4 This type of structure leads to ionic compounds having the following properties:

5 They are high melting point solids; multi-charged ions have greater electrostatic attractions and therefore higher melting points.

6 They do not conduct electricity in the solid state because the ions are fixed in place in the lattice and not free to move.

7 They do conduct electricity when molten or in aqueous solution because the ions are now free to move.

PRACTICE

1 $_{11}Na$ has 11 protons and thus 2.8.1 electrons. It therefore forms ions of symbol Na^+. What is the symbol of the ions of:

a) $_{19}K$ b) $_{9}F$ c) $_{3}Li$ d) $_{13}Al$ e) $_{16}S$

2 State one similarity and a difference between the magnesium and the sodium ion.

3 Draw dot and cross diagrams for the formation of calcium chloride from $_{20}Ca$ and $_{17}Cl$ atoms.

Covalent bonding

THE BARE BONES

- ➤ When atoms share electrons covalent bonds are formed.
- ➤ A covalent bond is a strong bond and is very difficult to break.
- ➤ Covalent substances exist as separate molecules or giant structures.

A Forming covalent bonds

1 When non-metals react together they both need to gain electrons and so they do this by sharing.

KEY FACT

2 Atoms are joined by shared pairs of electrons, with one electron from each atom making up the pair. The shared pair of electrons is a covalent bond.

3 The shared pair of electrons orbits around both atoms.

4 **Example:** the formation of methane (CH_4) from carbon and hydrogen atoms.

Each C atom has 4 outer electrons and each H atom has 1 outer electron. Sharing gives both of them full outer shells.

Remember
Atoms react in order to attain a complete outer shell.

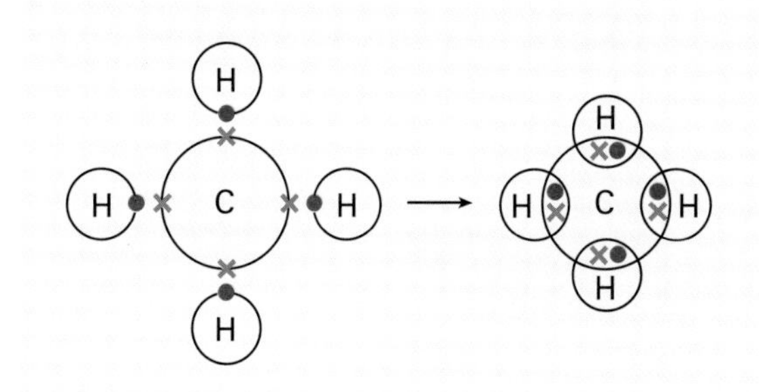

As only the outer electrons are used, only these are shown. This can also be shown as:

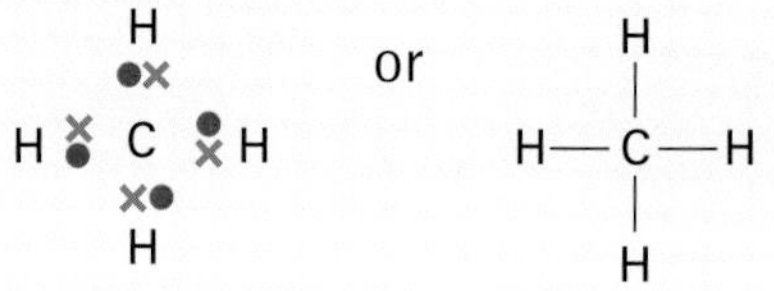

H
H ×● C ●× H
H

or

H—C—H (with H above and H below C)

Q What is a covalent bond?

B Other covalent molecules

Remember
In molecular oxygen, there are two shared pairs of electrons in the bond. This is called a double bond.

Water (H_2O):

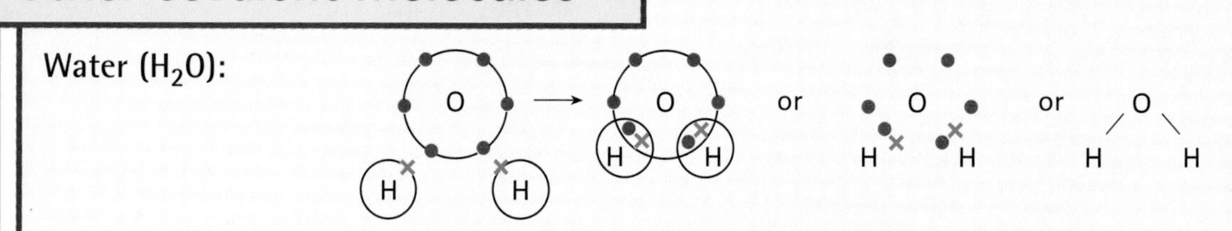

Ammonia (NH_3):

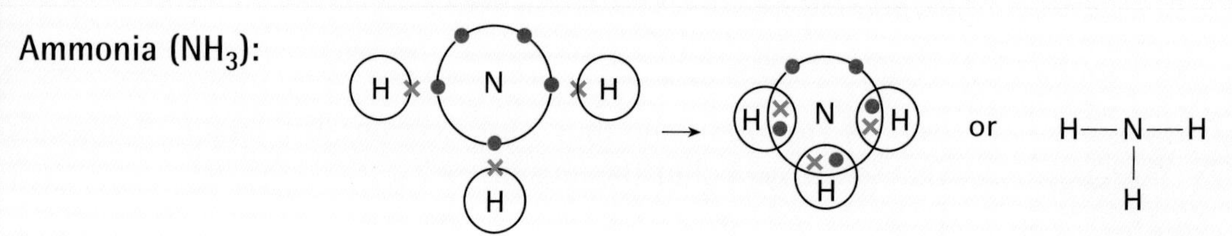

Oxygen (O_2):

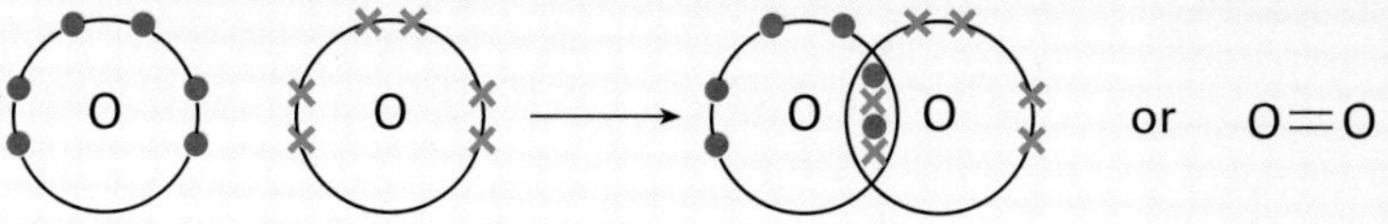

O O → O O or O=O

Q What is a 'double bond'? Now explain the term 'single bond'.

C Properties of covalent compounds

1 Most covalent compounds are made up of single, separate molecules.

KEY FACT

2 The covalent bonds within the molecules are very strong, but the forces between molecules are very low.

3 These compounds therefore have very low melting and boiling points and are often gases at room temperature and pressure.

4 Covalent molecules are neutral electrically and thus do not conduct electricity.

KEY FACT

5 Sometimes, covalent bonding forms giant structures, called macromolecules, with atoms bonded to each other continuously. The covalent bond can be so strong, that the giant structures are very hard solids with very high melting points.

6 **Example:** Diamond is pure carbon and each carbon atom forms four covalent bonds to four other carbon atoms in a rigid, giant structure. Diamond is the hardest of all natural substances.

You would not be expected to draw this diamond structure or the graphite structure below. You would be expected to be able to explain and talk about them.

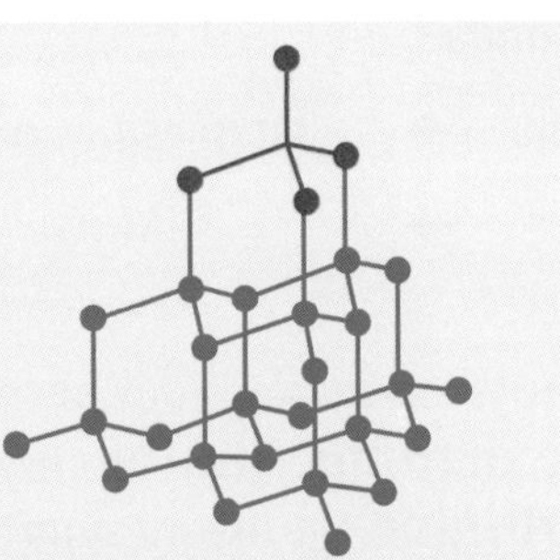

7 **Example:** Graphite is also pure carbon and has a layer structure. The carbon atoms form three covalent bonds and have one spare, unused electron:

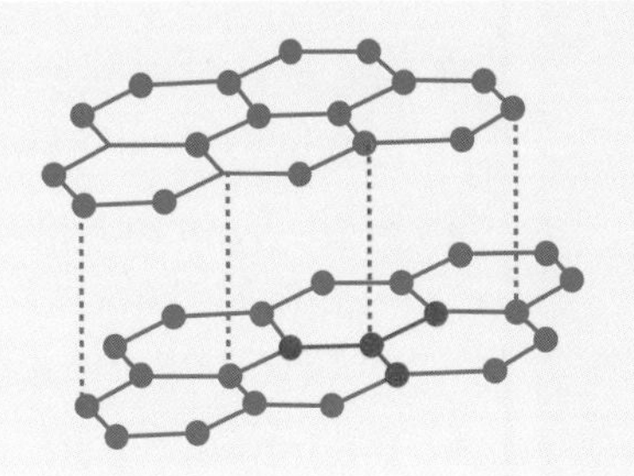

- The weak forces allow the layers to slide over each other, so graphite is soft and can be used as a lubricant and in pencils.
- The free electrons on each atom allow graphite to conduct electricity.

Q Explain why covalent compounds do not conduct electricity.

PRACTICE

1 Sketch dot and cross diagrams to show the following formulae:
a) H_2 from two $_1H$ atoms
b) HCl from $_1H$ and $_{17}Cl$ atoms

2 State and explain the main properties of simple molecular compounds.

3 Why do giant covalent structures (macromolecules) have high melting points?

Chemical equations

- All chemical reactions can be represented by a word equation and a balanced chemical equation.
- Reactions of ionic compounds have ionic equations.

Remember
In a chemical reaction the starting materials are the reactants and the materials present at the end are the products.

Remember
Balance one symbol at a time, reading left to right; put balancing numbers only at the front of a symbol or formula; do a final check once you think it is balanced.

Q What does the balanced chemical equation tell us that the word equation does not?

A Writing balanced chemical equations

1 An example is burning magnesium in oxygen. Magnesium oxide is produced. The **reactants** are **magnesium and oxygen**; the **product** is **magnesium oxide**.

2 The **word equation** is: magnesium + oxygen → magnesium oxide.
This tells us **what** happens but does not tell us about **quantities**.

3 For quantitative information, the **word equation must be translated into a balanced chemical equation**. The first step in this process is to change the words into symbols and formulae.

so magnesium + oxygen → magnesium oxide, becomes

$$Mg + O_2 \rightarrow MgO$$

Note that:
a) **elements have symbols**, so magnesium becomes Mg
b) **gaseous elements are diatomic**, (two atoms combined) so oxygen becomes O_2
c) **compounds have formulae**, so magnesium oxide becomes MgO.

4 The next step is to **balance this symbol equation**. This means that the number of each kind of atom must be **equal** on each side of the equation.

- In the symbol equation $Mg + O_2 \rightarrow MgO$, there is
  - one Mg on each side: balanced
  - two O on the left, one O on the right: **not balanced**
- The formula itself must **never** be changed but you can put **multiplying numbers** in front of a formula.
- You therefore multiply MgO by 2 to get $Mg + O_2 \rightarrow 2MgO$
- The O now balances with 2 on each side, but the Mg no longer balances as there is one on the left and two on the right.
- You therefore multiply the Mg on the left by 2 to give:
  $2Mg + O_2 \rightarrow 2MgO$, **which balances.**

5 The final step is to indicate the **physical state** of each material – whether gaseous (g), liquid (l), solid (s) or aqueous (aq).

The final equation is therefore: $2Mg(s) + O_2(g) \rightarrow 2MgO(s)$

B Ionic equations

KEY FACT

1 It is sometimes helpful in a chemical equation to write the ionic compounds in terms of the ions they contain.

2 Example: Burning sodium in chlorine to produce sodium chloride.

word equation:	sodium	+	chlorine	→	sodium chloride
symbol equation:	Na	+	Cl_2	→	NaCl
balanced chemical equation:	2Na(s)	+	Cl_2(g)	→	2NaCl(s)

Remember
All compounds containing metals are ionic.

3 Sodium chloride is an ionic compound made up of Na^+ and Cl^- ions and can be written as these ions in the equation:

ionic equation:	2Na(s)	+	Cl_2(g)	→	$2Na^+Cl^-$(s)
or, writing the ions separately	2Na(s)	+	Cl_2(g)	→	$2Na^+ + 2Cl^-$(s)

4 Sometimes, when you write the ionic compounds in terms of the ions they contain, you see that some ions appear on each side of the equation and can be 'cancelled'.

5 Example: Passing chlorine gas into aqueous potassium bromide solution.

word equation:	chlorine	+	potassium bromide	→	potassium chloride	+	bromine
symbol equation:	Cl_2	+	KBr	→	KCl	+	Br_2
balanced equation:	Cl_2(g)	+	2KBr(aq)	→	2KCl(aq)	+	Br_2(l)
ionic equation:	Cl_2(g)	+	$2K^+Br^-$(aq)	→	$2K^+Cl^-$(aq)	+	Br_2(l)

The K^+ ions appear on each side and have no part. They can be 'cancelled' or left out.

simplest equation:	Cl_2(g)	+	$2Br^-$(aq)	→	$2Cl^-$(aq)	+	Br_2(l)

Q Write the symbols of the ions that make up:
a) magnesium oxide and
b) potassium chloride.

PRACTICE

1 Write word equations and balanced chemical equations for each of the following reactions:

a) hydrogen burning in oxygen to form water

b) zinc reacting with steam to form zinc oxide and hydrogen

c) nitrogen and hydrogen combining to form ammonia

d) sodium reacting with water to form sodium hydroxide and hydrogen.

2 Translate the following word equations into i) a balanced chemical equations and ii) an ionic equation:

a) potassium + oxygen → potassium oxide

b) sodium hydroxide + magnesium chloride → magnesium hydroxide + sodium chloride

Calculations from equations

- Important quantities, such as relative molecular mass, can be calculated from balanced chemical equations.
- One mole of an element or compound is its relative atomic or molecular mass in grams.

A Calculating masses and percentage masses

1 Because a balanced chemical equation is **quantitative**, it can be used as the basis of various calculations.

- As long as the **relative atomic mass** (A_r) of each atom involved is known, the **relative molecular masses** and **percentage compositions** can be calculated.

KEY FACT

2 The relative molecular mass (M_r) of a compound is the sum of the relative atomic masses of all the atoms in one molecule of the compound.

3 Example:

- What is the M_r of sodium carbonate (Na_2CO_3) if the relative atomic masses are: C = 12, O = 16, Na = 23?

Answer: - formula is Na_2CO_3

- therefore $M_r = (2 \times 23) + 12 + (3 \times 16) = 106$

4 This can now be extended to find the percentage by mass of each of the elements in the compound.

5 In sodium carbonate, therefore:

- the percentage of sodium is $\frac{2 \times 23}{106} \times 100 = 43.4$
- the percentage of carbon is $\frac{12}{106} \times 100 = 11.3$
- the percentage of oxygen is $\frac{3 \times 16}{106} \times 100 = 45.3$

(total 100%)

Q Calculate M_r for CO_2 if A_r C = 12, A_r O = 16.

The relative atomic masses (M_r) of atoms are always given in calculations - you never have to learn them.

B The mole

KEY FACT

1 1 mole of an element is its relative atomic mass in grams and 1 mole of a compound is its relative molecular mass in grams.

2 Example: (H = 1, C = 12, O = 16, S = 32)

- the atomic mass of carbon = 12, therefore 1 mole of carbon = 12g
- the molecular mass of H_2 = 2, therefore 1 mole of hydrogen gas = 2g
- the molecular mass of CO_2 = 44, therefore 1 mole of carbon dioxide = 44g
- the molecular mass of H_2S = 34, therefore 1 mole of hydrogen sulphide = 34g

B

3 1 mole of an element always contains the same number of atoms: 6.02×10^{23}.

4 1 mole of a compound always contains the same number of molecules: 6.02×10^{23}.

5 6.02×10^{23} is the Avogadro number.

6 This means that:
- 12g of carbon contain 6.02×10^{23} C atoms
- 2g of hydrogen contain 6.02×10^{23} H_2 molecules
- 44g of carbon dioxide contain 6.02×10^{23} CO_2 molecules
- 34g of hydrogen sulphide contain 6.02×10^{23} H_2S molecules.

Q How many molecules are there in 36g of H_2O (H = 1, O = 16)?

C Reacting masses

Knowing the mass of any one of the reactants and products involved in a reaction enables the balanced chemical equation to be used to calculate the reacting mass of any or all of the other reactants and/or products.

Remember
Work precisely in columns of reactants and products mentioned in the question, under the equation.

Example 1:

- What mass (in grams) of magnesium oxide is produced by the complete combustion of 3g of magnesium in pure oxygen? (relative atomic masses: O = 16, Mg = 24)

Answer:

■ the balanced equation is:	2Mg	+ O_2	→	2MgO
■ put the masses under each:	2×24	2×16		$2 \times (24 + 16)$
■ the question is in grams so:	48 g	32 g		80 g
■ do nothing further in the oxygen column as you are not asked about it				
■ divide by 48 to get 1g of Mg:	1 g			$\frac{80}{48}$ g
■ multiply by 3 to get 3g of Mg:	3 g			$(\frac{80}{48} \times 3)$ g
■ therefore	3 g			5 g

■ 5 g of magnesium oxide is produced by the complete combustion of 3 g of magnesium.

Q How many grams are represented by 4NaOH in a chemical equation (H = 1, O = 16, Na = 23)?

PRACTICE

(Relative atomic masses: H = 1, C = 12, O = 16, Na = 23, Cu = 64)

1 In the reaction $2H_2O \rightarrow 2H_2 + O_2$, what mass of water is needed to produce 1g of hydrogen?

2 In the reaction $CuCO_3 \rightarrow CuO + CO_2$, what mass of copper oxide would be produced if 6.2g of copper carbonate are completely decomposed?

3 What mass of sodium must be burned completely in pure oxygen to produce 1.24g of sodium oxide?

Volume and formulae

- ➤ Pure compounds always contain elements combined in fixed proportions.
- ➤ The formula of a compound can be calculated from experiments that give the ratio by weight of its elements.
- ➤ With reactions involving gases, volumes are often easier to measure than weights.

A Reacting volumes

1 Because gases are so light and therefore difficult to weigh, it is often easier to measure the volume of a gas involved in a reaction.

KEY FACT

2 When stating the volume of a gas, the temperature and pressure at which that volume is measured must also be stated.

3 This is usually room temperature and pressure (r.t.p.), taken to be 20°C and 1 atm, or standard temperature and pressure (s.t.p.) which is 0°C and 1 atm.

KEY FACT

4 It is found that at r.t.p. the volume occupied by 1 mole of all gases is 24 dm^3 (24,000 cm^3). 24 dm^3 is the gas molar volume.

5 Example: (H = 1, C = 12, N = 14, O = 16)

- the volume of 2 g of H_2 at r.t.p. is 24 dm^3
- the volume of 44 g of CO_2 at r.t.p. is 24 dm^3
- the volume of 28 g of N_2 at r.t.p. is 24 dm^3
- the volume of 17 g of NH_3 at r.t.p. is 24 dm^3, and so on for all gases.

6 This fact enables you to make calculations about reacting volumes of gases.

7 Example: What volume of carbon dioxide is formed at r.t.p. from the complete combustion of 6 g of carbon (C = 12)?

Answer:

■ equation:	C	+	O_2	→	CO_2
■ masses and volumes:	12				24 dm^3
■ divide by 12:	1				$\frac{24}{12}$ dm^3
■ multiply by 6:	6				$\frac{24}{12} \times 6$ dm^3 = 12 dm^3

■ 12 dm^3 of CO_2 are formed at r.t.p. from the combustion of 6 g of carbon.

A

8 Example: At r.t.p. what volume of ammonia is produced when 11.2 dm^3 of hydrogen are completely reacted with nitrogen?

Answer:

- equation: N_2 + $3H_2$ → $2NH_3$
- volumes: 3×22.4 dm → $2 \times 22.4\ dm^3$
  - 67.2 dm^3 → 44.8 dm^3
- divide by 67.2: 1 dm^3 → $\frac{44.8}{67.2}\ dm^3$
- multiply by 11.2: 11.2 dm^3 → $\frac{44.8}{67.2} \times 11.2 = 7.5\ dm^3$
- 7.5 dm^3 of ammonia is produced from 11.2 dm^3 of hydrogen at r.t.p.

Q At r.t.p., what volume is occupied by
a) 3 moles of N_2
b) 5 moles of H_2S?

B Calculating formulae

1 Whenever a pure compound is made, it always contains the same elements in the same fixed proportions by weight.

2 This is called the Law of Constant Composition. It enables formulae to be calculated.

3 Example: 1.2 g of magnesium was burned to produce 2.0 g of magnesium oxide. Calculate the formula of magnesium oxide (O = 16, Mg = 24):

Answer:

- mass of magnesium oxide = 2.0 g
- mass of magnesium = 1.2 g
- therefore mass of oxygen used = 0.8 g
- ratio of magnesium: oxygen by mass = 1.2:0.8
- divide each by its A_r = $\frac{1.2}{24} : \frac{0.8}{16}$ = 0.05 : 0.05 = 1:1
- If magnesium reacts with oxygen in the ratio 1:1, then the formula of magnesium oxide must be MgO.

4 MgO is the simplest possible formula and is called the empirical formula.

Q What does the Law of constant composition state?

PRACTICE

1 Calculate the empirical formula of the compound formed when:

a) 7 g of nitrogen reacts with 16 g of oxygen.
b) 4.6 g of sodium reacts with 1.6 g of oxygen.

(C = 12, N = 14, O = 16, Na = 23: gas molar volume at r.t.p. is 24 dm^3)

Acids, bases and neutralisation

- All acids contain hydrogen ions (H^+) and all alkalis contain hydroxide ions (OH^-).
- The reaction between them to form water molecules (H_2O) is called neutralisation.

A Acids

1 Pure sulphuric acid consists mainly of covalent H_2SO_4 molecules and does not conduct electricity.

2 Sulphuric acid solution is an excellent conductor of electricity and must, therefore, contain ions.

3 Dissolving the H_2SO_4 molecules in water must cause them to ionise:

$H_2SO_4(l)$	→	$2H^+(aq)$	+	$SO_4^{2-}(aq)$
sulphuric acid		hydrogen ions		sulphate ion

KEY FACTS

4 The hydrogen ion is present in all acids and is responsible for the characteristic properties of acids.

5 An acid is a compound that produces hydrogen ions when in solution in water.

6 The more hydrogen ions an acid produces, the stronger it is and the lower its pH.

- **Example:** In hydrochloric acid (HCl) practically all the molecules ionise

$$HCl(aq) \xrightarrow[\text{water}]{\text{in}} H^+(aq) + Cl^-(aq);$$ HCl is a strong acid with pH about 1.

- **Example:** In carbonic acid (H_2CO_3) few molecules ionise

$$H_2CO_3(aq) \xrightarrow{\text{some}} 2H^+(aq) + CO_3^{2-}(aq)$$

and so carbonic acid has fewer H^+ ions, is a weak acid and pH around 5.

8 The pH scale goes from 0 to 14:

pH scale

0	1	2	3	4	5	6	7	8	9	10	11	12	13	14
strong acid					weak acid		neutral	weak alkali						strong alkali

Remember
Acid + Base → Salt + Water

Acid + Metal → Salt + Hydrogen

Acid + Carbonate → Salt + Water + Carbon Dioxide

Remember
Low pH: acid;
high pH: base

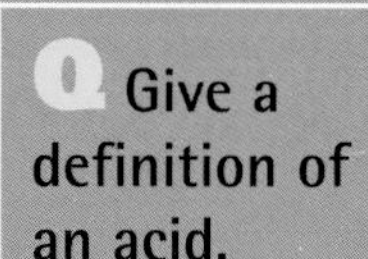
Give a definition of an acid.

B Bases

KEY FACT

1 **Bases are oxides and hydroxides of metals.**

2 Most of them are **insoluble in water**, but a few, notably **sodium hydroxide** (NaOH) and **potassium hydroxide** (KOH), are **soluble**.

3 On dissolving, they ionise to form hydroxide ions: $NaOH(aq) \xrightarrow[\text{water}]{\text{in}} Na^+(aq) + OH^-(aq)$

4 These **soluble bases** are called **alkalis**.

KEY FACT

5 **An alkali is a compound which ionises in water to produce hydroxide ions.**

6 A solution of NaOH or KOH is almost completely ionised so they are both **strong alkalis** of **pH around 14**.

7 In contrast, ammonium hydroxide solution **does not completely ionise**. It is a **weak alkali of about pH 10**.

$$NH_4OH(aq) \xrightarrow{\text{some}} NH_4^+(aq) + OH^-(aq)$$

Q Give a definition of an alkali.

C Neutralisation

1 The reaction between hydrochloric acid and soldium hydroxide is:

$$HCl(aq) + NaOH(aq) \rightarrow NaCl(aq) + H_2O(l)$$

2 As an ionic equation: $H^+Cl^-(aq) + Na^+OH^-(aq) \rightarrow Na^+Cl^-(aq) + H_2O(l)$

3 Being present on both sides, the Na+ and Cl- ions obviously take no part.

4 The **simplest ionic equation** is therefore: $H^+(aq) + OH^-(aq) \rightarrow H_2O(l)$

5 Any reaction between acid and alkali results in a solution of a salt being formed. This solution will have a **pH of 7** and be **neutral**. The reaction is therefore called **neutralisation**.

6 **Neutralisation is a reaction in which the hydrogen ions of an acid and the hydroxide ions of an alkali form water molecules:**

$$H^+(aq) + OH^-(aq) \rightarrow H_2O(l)$$

Q What is neutralisation?

PRACTICE

1 Explain, stating the approximate pH range of each and naming an example of each, what is meant by: a) strong acid b) strong alkali c) weak acid d) weak alkali.

2 Write i) a balanced chemical equation, ii) the simplest ionic equation for the following reaction: ammonium hydroxide and hydrochloric acid.

Redox and decomposition

THE BARE BONES

- Decomposition occurs when a compound is broken down into two or more simpler substances.
- Oxidation and reduction take place together in redox reactions.

A Decomposition

KEY FACT

1 When decomposition is brought about by heating the process is called <u>thermal decomposition</u>.

2 Examples

- If green copper carbonate powder is heated it decomposes easily into black copper oxide and carbon dioxide:

$$CuCO_3(s) \xrightarrow{\text{heat}} CuO(s) + CO_2(g)$$

- If marble chips (calcium carbonate) are strongly heated they decompose with some difficulty:

$$CuCO_3(s) \xrightarrow[\text{heat}]{\text{strong}} CO(s) + CO_2(g)$$

Remember
Reactive metals have stable compounds (see page 98).

KEY FACT

3 The $CaCO_3$ requires much more heat than the $CuCO_3$ because calcium is a much more reactive metal than copper and therefore its compounds are much more stable.

4 The very stable compounds of the most reactive metals cannot be decomposed by heat at all. They are usually decomposed by electrolysis when molten:

$$2NaCl(l) \xrightarrow{\text{electricity}} 2Na(s) + Cl_2(g)$$

Q What is thermal decomposition?

B Redox reactions

KEY FACTS

1 Both <u>gain of oxygen</u> and <u>loss of hydrogen</u> in a reaction are defined as <u>oxidation</u>.

2 Similarly, loss of oxygen and <u>gain of hydrogen</u> are <u>reduction</u>.

3 These processes take place simultaneously in redox reactions.

4 Examples:

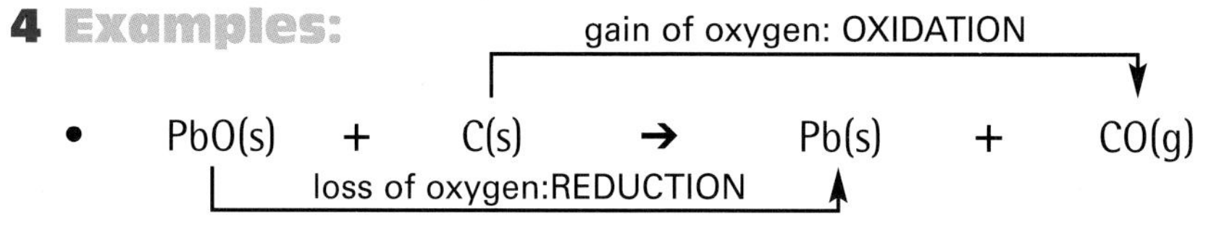

B

REDUCTION: gain of hydrogen (Cl_2 → $2HCl$)

- $H_2S(g) + Cl_2(g) \rightarrow 2HCl(g) + S(s)$

loss of hydrogen: OXIDATION (H_2S → S)

5 What actually takes place in redox reactions is the loss and gain of electrons.

KEY FACTS

6 An atom is said to be oxidised when it loses electrons.

7 An atom is said to be reduced when it gains electrons.

8 Example: $Zn(s) + 2HCl(aq) \rightarrow ZnCl_2(aq) + H_2(g)$

- The ionic equation is: $Zn(s) + 2H^+Cl^-(aq) \rightarrow Zn^{2+}2Cl^-(aq) + H_2(g)$
- Ignoring the Cl^- ions on both sides, the simplest ionic equation is:

$$Zn + 2H^+ \rightarrow Zn^{2+} + H_2$$

Remember
Acid + Metal → Salt + hydrogen

- The zinc atoms each lose two electrons (oxidation) and the $2H^+$ ions gain two electrons (reduction).

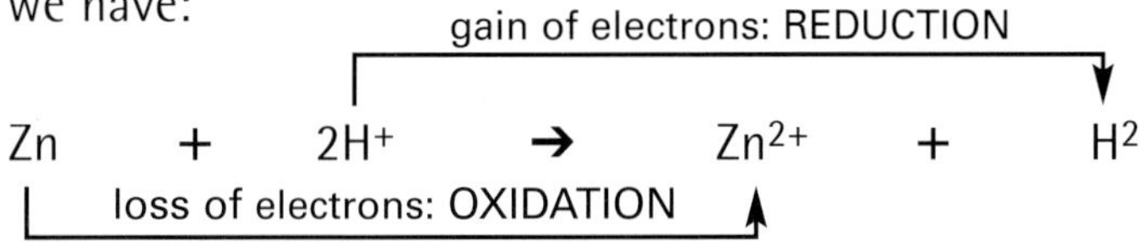

$Zn - 2e^- \rightarrow Zn^{2+}$: oxidation

$2H^+ + 2e^- \rightarrow H_2$: reduction

These are called electron-transfer, half-reaction equations — each of them represents half the full reaction.

- Overall we have:

gain of electrons: REDUCTION ($2H^+$ → H^2)

$$Zn + 2H^+ \rightarrow Zn^{2+} + H^2$$

loss of electrons: OXIDATION (Zn → Zn^{2+})

9 Looking again at the first example above: $PbO(s) + C(s) \rightarrow Pb(s) + CO(g)$

- the carbon is oxidised because it gains oxygen; the provider of the oxygen is the PbO; the PbO is the oxidising agent.
- the PbO is reduced because it loses oxygen; the taker of the oxygen is the C; the C is the reducing agent:

OXIDATION (C → CO)

$$PbO(s) + C(s) \rightarrow Pb(s) + CO(g)$$

oxidising agent: $PbO(s)$; reducing agent: $C(s)$

REDUCTION (PbO → Pb)

Q Give three definitions of oxidation.

10 So, in redox reactions, the substance that is oxidised is the reducing agent and the substance that is reduced is the oxidising agent.

PRACTICE

1 Put oxidation and reduction arrows on each of the following and say which substance is i) the oxidising agent, and ii) the reducing agent:

a) $CH_4(g) + 2Cl_2(g) \rightarrow C(s) + 4HCl(g)$

b) $CuO(s) + CO(g) \rightarrow Cu(s) + CO_2(g)$

2 Re-write $2Mg(s) + O_2(g) \rightarrow 2Mg^{2+}.O^{2-}(s)$ as two electron-transfer half-reaction equations, stating which is oxidation and which reduction and naming the oxidising agent and the reducing agent.

Hydrocarbons

- The chemistry of carbon compounds is called organic chemistry.
- Hydrocarbons are compounds which contain only carbon and hydrogen.
- They do combustion, substitution and addition reactions.

A Carbon and organic chemistry

Q What is a hydrocarbon?

1. Carbon atoms are the only atoms that can join to themselves, using covalent bonds, to form **chains**.
2. The chains can be up to thousands of atoms long.
3. The study of carbon compounds is called **organic chemistry**.

KEY FACT

4. The simplest organic compounds contain only carbon and hydrogen and thus are called <u>hydrocarbons</u>.

B Alkanes

1. The simplest hydrocarbons are those in which all the bonds are single bonds and are directed to different atoms. They are called alkanes.
2. The first five alkanes are listed in the table below:

Remember
You should know the difference between a molecular formula and a structural formula.

name	molecular formula	structural formula
methane	CH_4	H–C(H)(H)–H
ethane	C_2H_6	H–C(H)(H)–C(H)(H)–H
propane	C_3H_8	H–C(H)(H)–C(H)(H)–C(H)(H)–H
butane	C_4H_{10}	H–C(H)(H)–C(H)(H)–C(H)(H)–C(H)(H)–H

3. The **molecular formula** indicates how many of each kind of atom there is in one molecule and the **structural formula** shows the bonds between these atoms.
4. A group of compounds, such as the alkanes, which differ from each other by $^{-}CH_2$ and which have a **general formula** (C_nH_{2n+2} for alkanes) is called a **homologous series**.

Q The seventh alkane is heptane. State its molecular and structural formulas.

KEY FACT

5. Compounds like the alkanes which contain only <u>single bonds</u> are termed <u>saturated compounds</u>.

C Alkenes

KEY FACTS

1 Alkenes are a group of hydrocarbons containing double bonds – i.e. at least one pair of carbon atoms are joined by two shared pairs of electrons.

2 Compounds with double bonds are unsaturated compounds.

3 The first three alkenes in the series (general formula C_nH_{2n}) are:

ethene C_2H_4	propene C_3H_6	butene C_4H_8
$H_2C=CH_2$	$CH_3-CH=CH_2$	$CH_3-CH=CH-CH_3$

Q Explain the difference between saturated and unsaturated hydrocarbons.

D Reactions of alkanes and alkenes

1 **Combustion** is the most important reaction of **alkanes** – hence their use as fuels (see page 96):

$$2C_2H_6(g) + 7O_2(g) \rightarrow 4CO_2(g) + 6H_2O(g)$$

2 Saturated hydrocarbons undergo **substitution reactions**:

$$\underset{\text{ethane}}{C_2H_6(g)} + \underset{\text{chlorine}}{Cl_2(g)} \rightarrow \underset{\text{chloroethane}}{C_2H_5Cl(l)} + \underset{\text{hydrogen chloride}}{HCl(g)}$$

- A substitution reaction is one in which **one atom in a molecule is directly replaced by another atom.**

3 **Unsaturated** hydrocarbons, such as alkenes, undergo **addition reactions** to form saturated compounds

molecular equation: $\underset{\text{ethene}}{C_2H_4(g)} + H_2(g) \rightarrow \underset{\text{ethane}}{C_2H_6(g)}$

structural equation: $H_2C=CH_2 + H-H \rightarrow H_3C-CH_3$

- An addition reaction is one in which **a small molecule adds on across the double bond** of an unsaturated molecule.

Q Give a molecular and structural formula for the fourth alkene, pentene.

PRACTICE

1 Draw the structural formula of: a) ethane b) propene c) chloropropane

2 What is: a) a substitution reaction b) an addition reaction?
Give an example of each, showing the structural formulae involved.

3 Give a molecular equation and a structural equation for:
a) the reaction between butane and chlorine
b) the reaction between propene and hydrogen.

Crude oil, petrol and polythene

- Crude oil is the source of many chemicals used as fuels, and to make plastics.
- Distillation and cracking can be applied to crude oil to obtain useful products.
- Polymers are compounds, such as plastics, made by joining together small molecules into long chains.

A Crude oil and its distillation

1 Crude oil is a fossil fuel formed by the action of heat and pressure over millions of years on plant and animal remains.

KEY FACT

2 It is a mixture of hydrocarbons with chains of various lengths which can be separated by fractional distillation into fractions, each containing molecules with a similar number of carbon atoms.

Remember
Hydrocarbons contain only hydrogen and carbon.

3 By evaporating crude oil and allowing it to condense at different temperatures, a number of different fractions can be obtained.

4 The following table shows some of these fractions:

approx temp (°C)	name of fraction	approx. no of C atoms	uses
20	gas	1-4	simple alkanes for fuels
70	naphtha	5-10	petrol for vehicle fuel
150	paraffin	10-16	aviation and heater fuel, detergents
240	diesel	16-20	diesel fuel, chemicals
350	lubricating oil/wax	20-60	waxes, greases, ship and power station fuel
500	bitumen	60 plus	roads, roofing, waterproofing

You would not be expected to learn this table.

KEY FACT

5 Most of the products are used as fuels because all hydrocarbons burn easily in air.

$C_3H_8(g)$	+	$5O_2(g)$	→	$3CO_2(g)$	+	$4H_2O(g)$	+	heat
propane		oxygen		carbon dioxide		water		

Q What is fractional distillation?

B Cracking

1 There is a much greater demand for **short-chain** than long-chain fractions.

2 Long-chain hydrocarbons can be broken down into smaller chains by passing the vapour over a **heated catalyst**, such as aluminium oxide. A **thermal decomposition reaction occurs.** This breaking down is called **cracking**.

3 Example:

$$C_{19}H_{40} \xrightarrow[\text{catalyst}]{\text{heat}} C_{10}H_{22} + C_4H_8 + C_3H_6 + C_2H_4$$

$C_{10}H_{22}$ for petrol; C_4H_8, C_3H_6, C_2H_4 alkenes for polymerisation (see below)

Q What is cracking and why is it so useful?

C Polymerisation

KEY FACT

1 **The double bond in ethene enables many ethene molecules to add together to form a single, long-chain molecule.**

$$\cdots \; CH_2{=}CH_2 + CH_2{=}CH_2 + CH_2{=}CH_2 + \cdots \xrightarrow[\text{catalyst}]{\text{heat, pressure}} \cdots -CH_2-CH_2-CH_2-CH_2-CH_2-CH_2- \cdots$$

2 The **structural equation** above is more easily written as a **general molecular equation**:

$$n\,CH_2 = CH_2 \xrightarrow[\text{catalyst}]{\text{heat, pressure}} \left(CH_2 - CH_2 \right)_n$$

ethene → poly(ethene)

KEY FACT

3 **The giant molecule produced is a <u>polymer</u>; the individual ethane molecules are <u>monomers</u>; the process is <u>polymerisation</u>; polymers produced like this, by <u>adding</u> monomers together, are <u>addition polymers</u> and the reaction is <u>addition polymerisation</u>.**

4 Poly(ethane) is better known as **polythene** and is used for plastic bags, bottles, kitchenware etc.

5 The **advantages** of these polymers (sometimes referred to as **plastics**) is that they are **flexible, strong, non-corrosive and cheap**.

6 The greatest **disadvantage** is that they cannot be broken down by micro-organisms and therefore they are **not biodegradable**.

Q What is polymerisation?

PRACTICE

1 Write a balanced chemical equation for the combustion of:
a) ethane b) propane

2 Vinyl chloride (C_2H_3Cl) has the structural formula: $H_2C{=}CH_2$ (H, H / C=C / H, H)
Write a general molecular equation, for its polymerisation into poly(vinyl chloride) [PVC].

Reactivity of metals

THE BARE BONES

- Metals can be arranged in the order of their reactivity.
- Reactive metals form compounds more easily and quickly than less reactive metals.
- As a result, reactive metals are more difficult to extract from their ores.
- Methods of extraction include reduction and electrolysis.

A The reactivity series

1 The reactivity series is a list of metals arranged in order of reactivity.

KEY FACT

2 The method used to extract a metal from its ore depends on the position of the metal in the reactivity series.

3 This is shown in the following table:

	metal	method of extraction	
most reactive ↑	potassium sodium calcium magnesium aluminium	electrolysis (see page 100)	most difficult to extract ↓
	zinc iron lead copper mercury	reduction of the oxide (see page 99)	
least reactive	silver gold	found naturally	least difficult to extract

Remember
Metals + :

- air/oxygen → oxide
- water/steam → hydroxide/oxide
- acid → salt + hydrogen

Q What is the 'reactivity series'?

You do not need to learn the order of metals, but you would be expected to use information given to make predictions about a metal's extraction and reactions.

B Displacement

KEY FACT

1 Metals can be displaced from their salts or oxides by a metal higher in the reactivity series.

2 Examples:

$$Zn(s) + CuSO_4(aq) \rightarrow ZnSO_4(aq) + Cu(s)$$

$$2Al(s) + Fe_2O_3(s) \rightarrow 2Fe(s) + Al_2O_3(s)$$

3 But, $Al(s) + MgCl_2(aq) \nrightarrow$ no reaction because Al is less reactive than Mg.

Q Why will zinc displace copper from its salts but not magnesium from its salts?

C Extraction by reduction

KEY FACTS

1 Many necessary materials are obtained from the rocks forming the Earth's crust: glass, hydrogen, chlorine, acids, ceramics, building materials, etc.

2 Rocks composed mainly of the compounds of one metal and from which we can extract the metal are called ores.

3 As outlined above, the less reactive metals are obtained by reduction of the oxide.

4 **Example:** The main ore of **iron** is haematite (mainly iron (III) oxide). It is reduced to iron by carbon monoxide in the blast furnace:

$$Fe_2O_3(s) + 3CO(g) \rightarrow 2Fe(s) + 3CO_2(g)$$

5 **Example:** The main ore of **zinc** is zinc blende (mainly zinc sulphide). This is roasted in air to convert it to zinc oxide, which is then reduced by roasting with coke (carbon):

$$2ZnS(s) + 3O_2(g) \rightarrow 2ZnO(s) + 2SO_2(g)$$
$$2ZnO(s) + C(s) \rightarrow 2Zn(s) + CO_2(g)$$

6 **Example:** The main ore of **copper** is copper pyrite ($CuFeS_2$). Roasting this in air converts it first to copper (I) sulphide (Cu_2S) and then reduces this to copper:

$$2\ CuFeS_2(s) + 4O_2(g) \rightarrow Cu_2S(s) + 2FeO(s) + 3SO_2(g)$$
$$Cu_2S(s) + O_2(g) \rightarrow 2Cu(s) + SO_2(g)$$

The copper obtained is about 98% pure and is called 'blister copper'.

It is further purified **electrolytically**.

It is likely that you have studied at least one of these examples in some detail and you should be familiar with these details.

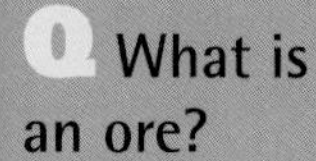

Q What is an ore?

PRACTICE

1 Potassium is extracted from its ore by electrolysis. What does this tell you about potassium?

2 For each of the following pairs, give the balanced chemical equation if you predict they will react, and write 'no reaçtion' if you predict otherwise, giving a reason for there being no reaction:

a) magnesium and zinc sulphate
b) copper and magnesium chloride
c) zinc and copper oxide
d) magnesium and copper nitrate

3 Mercury can be obtained by the thermal decomposition of its oxide (HgO).

a) write a balanced chemical equation for this reaction
b) explain why zinc cannot be obtained from zinc oxide in this way.

Electrolysis

- Electrolysis is the breaking down of ionic substances using an electric current.
- Electrolysis is used commercially to extract reactive metals from their ores.

A General principles

1 When an ionic substance is molten or in solution the ions are free to move.

KEY FACT

2 In electrolysis, positively charged ions are attracted to the cathode (–), and negatively charged ions to the anode (+). Here, the ions may undergo chemical reactions by gaining or losing electrons, respectively.

B Extraction of sodium

Remember
Electron loss is oxidation and electron gain is reduction.

Q Why does molten chloride conduct electricity when solid sodium chloride does not?

1 Sodium is a very reactive metal. It is obtained by the electrolysis of molten sodium chloride (NaCl). In the melt, the ions present are Na^+ and Cl^-.

2 At the anode(+): negative ions are attracted to the positive anode, hence Cl^- ions are attracted and lose electrons.
$2Cl^- - 2e^- \rightarrow Cl_2(g)$: oxidation.

3 At the cathode(–): positive ions are attracted to the negative cathode, hence Na^+ ions are attracted and gain electrons.
$2Na^+ + 2e^- \rightarrow 2Na(s)$: reduction.

4 The overall reaction is:

$$2NaCl(l) \xrightarrow[\text{decomposition}]{\text{electrolytic}} 2Na(s) + Cl_2(g)$$

At the electrodes: the negative (minus) ions lose (minus) electrons; the positive (plus) ions gain (plus) electrons.

C Extraction of aluminium

1 The high reactivity of aluminium means it also needs to be extracted by electrolysis.

2 The main ore of aluminium is bauxite (mainly Al_2O_3).

KEY FACTS

3 Pure Al_2O_3 has a melting point of 2400°C and electrolysis at this temperature is both expensive and dangerous.

4 Purified Al_2O_3 is therefore dissolved in molten cryolite (Na_3AlF_6) which melts at 800°C, the temperature at which the electrolysis is carried out.

C

KEY FACT

5 In the cryolite the Al^{3+} and O^{2-} ions in the Al_2O_3 become free to move as the lattice breaks down. The cryolite acts only as a solvent and takes no part in the chemical process, which is:

Ions present:	$Al_2O_3 \rightarrow 2Al^{3+} + 3O^{2-}$
At cathode (−):	Al^{3+} ions attracted
	$2Al^{3+} + 6e^- \rightarrow 2Al(s)$: reduction
At anode (+):	O^{2-} ions attracted
	$3O^{2-} - 6e^- \rightarrow 3O$: oxidation
then,	$3O + 3O \rightarrow 3O_2(g)$
Overall:	$2Al_2O_3 \xrightarrow[\text{decomposition}]{\text{electrolytic}} 4Al(s) + 3O_2(g)$

Q Why is cryolite used in the extraction of aluminium from bauxite?

D Purification of copper

1 Copper extracted from its ore is 98% pure and called 'blister' copper.

KEY FACT

2 A lump of blister copper is used as the anode of a cell in which the cathode is a strip of pure copper and the electrolyte is copper (II) sulphate solution.

Ions present:	$CuSO_4 \rightarrow Cu^{2+} + SO_4^{2-}$
At cathode (−):	Cu^{2+} ions are attracted
	$Cu^{2+} + 2e^- \rightarrow Cu(s)$: reduction
At anode (+):	The copper atoms of the blister copper each lose two electrons and dissolve into the solution.
	$Cu - 2e^- \rightarrow Cu^{2+}(aq)$: oxidation
Overall:	The copper that dissolves out of the impure blister copper is deposited as pure copper on the cathode.

Remember
The most reactive metals have the most stable compounds and therefore are the most difficult to extract.

Q What is 'blister' copper?

PRACTICE

1 Copy and complete the following half-reaction equations:
a) anode: $2I^- - 2e^- \rightarrow$ ....
b) anode: $2O^{2-} -$ .... $\rightarrow O_2(g)$
c) cathode: .... $+ 2e^- \rightarrow 2K(s)$

2 Explain, with equations, the electrolysis of molten magnesium chloride ($MgCl_2$).

3 Explain why the electrolytic decomposition of aluminium oxide is a redox reaction.

Ammonia and fertilisers

- Nitrogen from air is used to manufacture ammonia.
- Ammonia is used to make nitrogen-based fertilisers.
- Fertilisers can cause problems if they get into streams, rivers or drinking water.

A Manufacture of ammonia

Remember
Air is about 80% nitrogen. The nitrogen is unreactive so using it in reactions is difficult.

Ammonia is manufactured by the **Haber process**. The raw materials are **natural gas** (CH_4, methane), **air** and **water**. (Air is about 80% nitrogen.) The five stages of the process are:

- $$CH_4(g) + 2H_2O(g) \xrightarrow[\text{Ni catalyst}]{300\text{ atm}/900^\circ C} CO_2(g) + 4H_2(g)$$
- Air (O_2 and N_2) is added and some of the hydrogen produced above converts the oxygen to steam:
  $$2H_2(g) + O_2(g) \rightarrow 2H_2O(g)$$
- The mixture is now CO_2, steam, N_2, H_2. Passing this through concentrated potassium hydroxide solution removes CO_2 and steam.
- The N_2 and H_2 remaining are reacted together:
  $$N_2(g) + 3H_2(g) \xrightarrow[\text{Fe catalyst}]{250\text{ atm}/550^\circ C} 2NH_3(g)$$
- This reaction is **reversible** and **exothermic** [see page 120 for further details]. About 15% of the gases are converted to ammonia, which is removed as a liquid by pressure and cooling. The unreacted N_2/H_2 are recycled to make more ammonia.

Q What is the importance of manufacturing ammonia?

B Making nitrogenous fertilisers

1 A concentrated solution of ammonia can itself be used as a fertiliser, but ammonium salts are better as they can be stored as solids.

KEY FACT

2 **The ammonium salts are made by neutralising the ammonia solution (an alkali) with acid.**

They form, for example, ammonium sulphate:

$$2NH_3(aq) + H_2SO_4(aq) \rightarrow (NH_4)_2SO_4(aq)$$

or ammonium nitrate:

$$NH_3(aq) + HNO_3(aq) \rightarrow NH_4NO_3(aq)$$

both important nitrogenous fertilisers

Q What is meant by 'nitrogenous fertilisers'?

C Fertilisers and plants

KEY FACT

1 Plants take several kinds of nutrients from the soil. Nutrients containing nitrogen are needed for plants to make proteins.

2 These nutrients must be replaced before further crops can be grown.

KEY FACT

3 Nutrients can be replaced by natural fertilisers, such as manure and compost, but the economic and social pressure on farmers to produce high yields of healthy, sizeable crops very quickly means that artificial fertilisers are required.

Q What is the difference between 'natural' and 'artificial' fertilisers?

4 Artificial fertilisers contain elements such as **nitrogen (N), phosphorus (P) and potassium (K)**. Fertilisers containing all of these are sometimes referred to as **NPK fertilisers**.

D Fertilisers and pollution

1 When fertilisers wash into lakes and rivers, the natural cycles are upset.

KEY FACT

2 The fertiliser causes algae to multiply; when the algae die the bacteria that feed on dead algae multiply; the increased bacteria consume increased amounts of oxygen; fish die from lack of oxygen.

Q How might nitrates get into drinking water?

3 Nitrates in drinking water are converted into **nitrites** (the NO_2^- ion). Nitrites can affect health by **oxidising the iron in haemoglobin**, thus preventing the haemoglobin from carrying oxygen around the body as it ought to do. Babies can turn blue from lack of oxygen in this way.

1 Give a balanced equation, and state the conditions, for the conversion of nitrogen and hydrogen into ammonia in the Haber process.

2 What functions do fertilisers perform in soil and what effect do they have on plant growth?

3 Why do some lakes develop a surface layer of algae, and why is this less likely to occur in a river?

4 What problems do layers of algae cause on a lake's surface?

The atmosphere

- The Earth's atmosphere has been chemically the same for about 200 million years.
- Up to that time, it was evolving and changing.
- The atmosphere is maintained as it is by natural cycles such as the carbon cycle.

Remember
Although the atmosphere is now mainly nitrogen (78%) and oxygen (21%), it has not always been like this.

A The changing atmosphere

1 4500 MILLION (m) YEARS AGO (approx.)

Earth first formed. Volcanic activity over the next 1000 million years put water vapour, CO_2, methane (CH_4), SO_2 into the atmosphere. Light gases (H_2, He) floated off into space.

2 3800m YEARS AGO

The water vapour condensed to form oceans and much CO_2 dissolved into them, eventually forming carbonates such as limestone.

3 3500m YEARS AGO

Simple plant life appeared. As this spread across the surface of the Earth:

- plants photosynthesised, removing CO_2 from the atmosphere and releasing O_2:

$$CO_2(g) + H_2O(g) \xrightarrow{\text{light}} \text{organic material} + O_2(g)$$

- the methane and ammonia in the atmosphere reacted with the oxygen, releasing CO_2 and N_2:

$$CH_4(g) + 2O_2(g) \rightarrow CO_2(g) + 2H_2O(g)$$

$$4NH_3(g) + 3O_2(g) \rightarrow 2N_2(g) + 6H_2O(g)$$

- living organisms, such as denitrifying bacteria, released N_2 into the atmosphere, adding to that produced from ammonia, above.

4 800m YEARS AGO

Enough oxygen had accumulated to enable an ozone layer to develop:

$$3O_2(g) \rightarrow 2O_3(g) \text{ (ozone)}$$

The ozone layer filters out harmful ultraviolet radiation from the Sun, allowing new and more complex living things to develop. Animal life thus began over 500m years ago. Animals using O_2 for respiration and plants using CO_2 for photosynthesis began to maintain a CO_2/O_2 balance, so that by about:

5 200m YEARS AGO

The atmosphere had become what it is today –
78% N_2; 21% O_2; 1% noble gases; 0.03% CO_2; varying % H_2O vapour.

Q How was oxygen first added to the atmosphere?

B The carbon cycle

1 Nature achieves balance in the atmosphere by recycling.

2 The carbon cycle is one example of this. Others are the nitrogen, oxygen and water cycles.

KEY FACT

3 The carbon cycle keeps the CO_2 in the atmosphere in balance at 0.03%.

You won't have to draw the carbon cycle, but you will probably have to answer questions on a diagram of it.

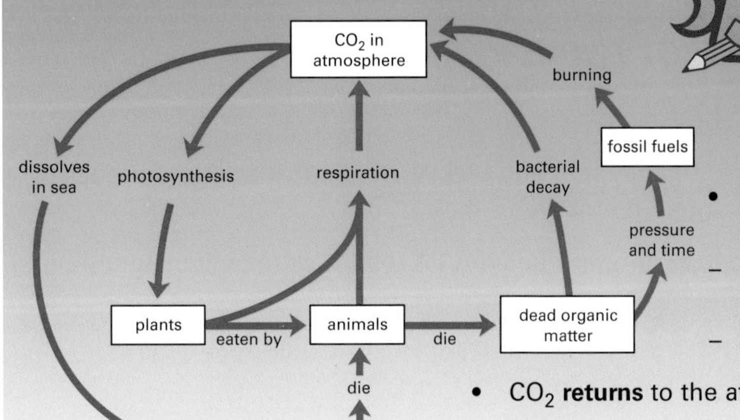

- CO_2 **leaves** the atmosphere by two routes:
  - CO_2 being absorbed into plants by photosynthesis
  - CO_2 dissolving into the sea.
- CO_2 **returns** to the atmosphere by three routes:
  - plants and animals releasing CO_2 during respiration
  - bacteria releasing CO_2 by feeding on dead plants and animals
  - dead plants and animals becoming fossil fuels which release CO_2 when burned.

KEY FACTS

4 At the present time, humans are burning an increasing amount of fossil fuel. This increases CO_2 in the atmosphere. At the same time we are cutting down large swathes of the great rainforests of the world. This removes plants which absorb CO_2.

5 Therefore, the CO_2 in the atmosphere is increasing. This heavy gas remains in the atmosphere and traps heat.

6 This greenhouse effect is raising the temperature of the Earth (global warming).

Q Why does nature recycle material?

PRACTICE

1 Outline the causes of:
a) the oceans forming; b) oxygen entering the atmosphere.

2 What key factor allowed complex animal life to begin on Earth?

3 Explain two ways in which carbon is returned to the atmosphere.

4 What do you think some of the effects of global warming might be and explain the problems each effect might cause?

Rocks

- The three main types of rock are igneous, sedimentary and metamorphic.
- Studying rocks gives clues to their formation and past history.
- The Earth's surface layers are divided into huge rigid 'plates' that move relative to each other.

A Structure of the Earth

The Earth consists of three layers:

1. The core (mainly iron and nickel) which divides into an inner core of solid rock and an outer core of liquid rock.
2. The mantle, around the core, of solid rock (mainly magnesium and silicon) which can move very slowly.
3. The crust – a thin (50km deep) layer of solid rock. (The following sections are largely concerned with crustal rocks.)

Q Name the three layers that make up the Earth.

B The three main rock types

KEY FACTS

1. Igneous rock is formed when molten rock (magma) rises and solidifies.

If the magma reaches the Earth's surface and erupts out of volcanoes it is called lava. This crystallises quickly (days/weeks) to form rocks with small crystals called extrusive igneous rock (e.g. basalt).

If the magma stops below the surface, it cools slowly (thousands of years) to form large crystals and is called intrusive igneous rock (e.g. granite).

Remember
On surface rapid cooling - small crystals - extrusive rock.
Below surface slow cooling - large crystals - intrusive rock.

2. Sedimentary rocks (e.g. limestone, coal and sandstone) are formed from compacted layers of sediment.

- Surface rocks are broken into sediment by weathering and erosion.
- The sediment is transported by wind, water, animals to be deposited in layers.
- The layers are compacted by pressure as layer upon layer is deposited.
- The oldest layers are the deepest.

KEY FACT

3. Metamorphic rocks are formed when igneous or sedimentary rocks are changed by heat/pressure, deep underground, into harder rock.

B

4 The **rock cycle** shows how rocks **change from one type to another.**

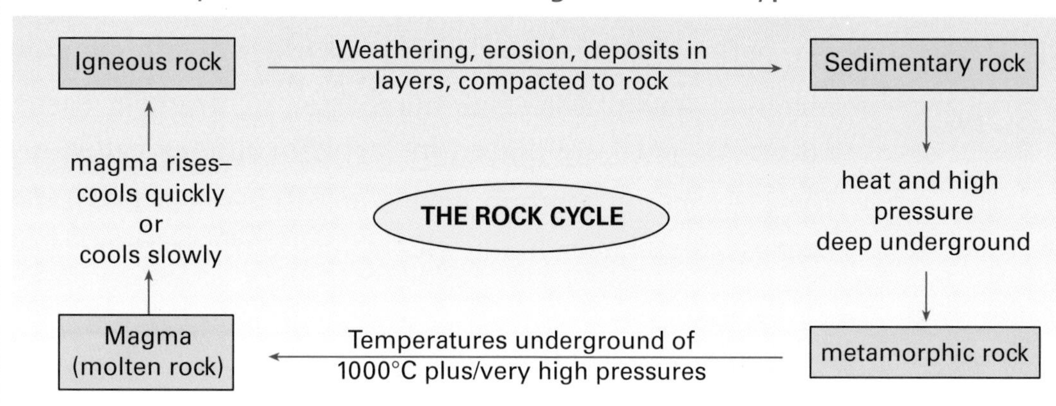

5 When rocks are broken up by **wind, rain or extremes of temperature**, this is called **physical weathering**. The effects of phenomena such as **acid rain** are called **chemical weathering**, while breaking by **plant roots, animals walking** etc., is called **biological weathering**.

Q Why do you think there was only one type of rock present when the earth first formed?

C The rock record

KEY FACT

1 Studying rocks suggests evidence for their formation and change.

2 Some examples of this are:

- Sedimentary rocks are often **tilted, folded or broken**, showing that movement must have occurred due to **enormous forces.**
- **Fossils** discovered in sedimentary layers suggest **age** and **formation time.**
- **Mountain ranges** are **formed** by large-scale **movement of the crust**, whilst older mountains are worn away by weathering and erosion.
- The **edges** of land masses (e.g. the Eastern coasts of N and S America and the Western Europe/Africa coastline) have **shapes** which appear to **fit closely** and have similar fossils and rocks, suggesting they were **once the same** land mass but have **moved apart.**
- Earthquakes and volcanoes show that **rock movement continues today** and information is obtained from these by studying shock waves.

Q How do new mountain ranges replace old ones?

PRACTICE

1 State the differences between basalt and granite and explain how they arose.

2 How can fossils help to tell us the ages of sedimentary rocks?

3 What evidence is there that land masses have moved apart?

The periodic table

- The periodic table lists elements in order of atomic number.
- Elements having the same number of electrons in their outermost shell are placed in vertical columns called groups.
- The elements in a group have similar chemical properties because they have similar electronic configurations.

A About the periodic table

1 The periodic table has 8 groups of elements plus a block of transition metals - slightly over 100 elements in total.

KEY FACTS

2 Elements having the same number of outermost electrons lie in the same group and have similar chemical properties.

3 With the exception of the noble gases (Group 0 - see page 116), the group number is the number of outermost electrons.

4 The horizontal rows are periods. Period 1 is hydrogen and helium; Period 2, lithium to neon; Period 3, sodium to argon; and so on. Elements in a period have the same electronic configuration in their inner shells.

KEY FACT

5 The transition metals have electronic configurations in which the penultimate (one-from-outermost) shell is being filled (see 'electronic configuration' on page 78 and 'transition metals' on page 112).

6 Moving left to right across a period shows a gradual change from the most metallic elements (Group 1 – the alkali metals) to the most non-metallic (Group VII – the halogens).

KEY FACT

7 Moving down a metal group sees a gradual increase in metallic properties, whilst moving up a non-metal group sees a gradual increase in non-metallic properties.

8 Thus, the most metallic elements are bottom of Group 1 and the most non-metallic are top of Group VII.

9 The zigzag line through the table separates the metals and non-metals. The non-metals are to the right of the line. There are 22 non-metals. All the other elements in the periodic table are metallic.

The full periodic table appears on page 117.

Remember
The noble gases of Group 0 are an exceptional case - see page 116.

Q What is the periodic table?

The grid below shows the position in the periodic table of eight unknown elements. (The letters are not the symbols of the elements.) Copy the grid and answer the questions below.

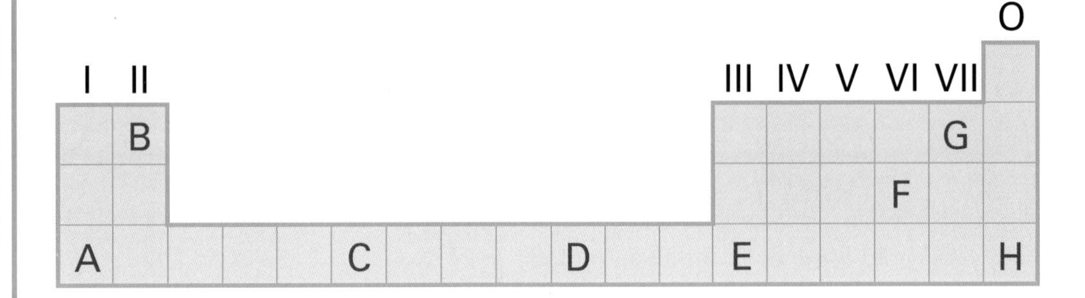

1 Give the letters of three metals in the grid.

2 Which element in the grid is the most metallic?

3 Give the letters of three non-metals in the grid.

4 Which element in the grid is the most non-metallic?

5 Which element reacts most easily with A?

6 What do elements C and D have in common?

7 X, Y, Z are three elements. From the information below, place them in the grid in a place they could be found in the periodic table.
a) element X is a halogen
b) element Y is the least metallic alkali metal
c) element Z has eight electrons in its outermost shell

8 J and K are in Group IV; J is a non-metal whilst K is a metal. Place J and K in the grid.

9 L has an atomic number of 29. Place it correctly in the grid. What type of element is it?

10 Place the following in an appropriate place in the grid:
a) M has electronic configuration 2.8.3
b) P has a complete outer shell of two electrons
c) R reacts with oxygen to form the ionic compound $R^{2+}O^{2-}$
d) T forms the T^{2-} ion in some of its reactions.

Group I – the alkali metals

- The alkali metals make up Group I of the periodic table, and include lithium, sodium and potassium.
- They are all very reactive.

A Properties

1 Basic data for alkali metals:

name	symbol	atomic number	electronic configuration	density (g/cm^3)	melting point (°C)	boiling point (°C)
lithium	Li	3	2.1	0.53	180	1336
sodium	Na	11	2.8.1	0.97	98	883
potassium	K	19	2.8.8.1	0.86	63.5	757
rubidium	Rb	37	2.8.18.8.1	1.53	39	697
caesium	Cs	55	2.8.18.18.8.1	1.88	29	670
francium	Fr	87	2.8.18.32.18.8.1	?	?	?
				increase ↓	decrease ↓	decrease ↓

Q Why does the melting point of the alkali metals decrease with increasing atomic number?

2 Physical properties of the metals: soft; shiny when cut; **low densities, increasing down the group** as the atom gets bigger, Li, Na, K have densities less than water (which is $1.0 g/cm^3$) and therefore they float on water.

3 **Melting points** and **boiling points** are **low** and **decrease down the group** as the inter-atomic forces are weaker in bigger atoms and therefore the atoms separate from their structure more easily on heating.

B Reactivity

KEY FACTS

1 All are very reactive and react by the loss of the single outer electron to form a unipositive ion: $M - e^- \rightarrow M^+$

2 Reactivity increases down the group as the atom gets larger and the single electron is lost more easily.

thus $Rb - e^- \rightarrow Rb^+$ (2.8.18.8.1 → 2.8.18.8) **is much easier than** $Li - e^- \rightarrow Li^+$ (2.1 → 2)

- This is because:

  i) the single outer electron is further from the nucleus and hence the force of its attraction to the nucleus lessens in the larger atom.

  ii) moving down the group, the single outer electron has more shells of inner electrons shielding it from the attraction of the positive nucleus, thus making it easier to remove.

Q What is a 'unipositive' ion?

C Typical reactions

The alkali metals react with:

- oxygen, by combustion, to form oxides:
  $$4M(s) + O_2(g) \rightarrow 2M_2O(s) \quad [(M^+)_2.O^{2-}]$$
- water, to produce an aqueous solution of the hydroxide – a strong alkali – and hydrogen:
  $$2M(s) + 2H_2O(l) \rightarrow 2MOH(aq) \quad [M^+.OH^-] \quad + H_2(g)$$
- the Group VII halogens to form halides:
  $$2M(s) + X_2(g) \rightarrow 2MX(s) \quad [M^+.X^-]$$

Q Why are the reactions of the alkali metals the same throughout the group?

D Alkali metal compounds

KEY FACTS

1 Alkali metals are very reactive and so have stable compounds (see Section A on page 98) which do not decompose on heating.

2 Alkali metals cannot be extracted from their compounds by chemical means, such as reduction of the oxide; electrolytic decomposition is needed.

$$2NaCl(l) \xrightarrow{\text{electrolysis}} 2Na(s) + Cl_2(g)$$

Remember
The OH^- ion is responsible for the characteristic properties of alkalis.

3 The chemical properties of alkali metal compounds can be seen as the chemical properties of the ions they contain:

Examples:

- All alkali metal compounds are soluble in water because water molecules break down the ionic lattices

Q Why are all alkali metal compounds water soluble?

KEY FACT

The hydroxides in aqueous solution are strong alkalis because they ionise completely to produce hydroxide (OH^-) ions.

$$NaOH(s) \xrightarrow{H_2O} NaOH(aq) \rightarrow Na^+(aq) + OH^-(aq)$$

- Formation of these strong alkalis gives the group the name 'alkali metals'.

PRACTICE

1 Francium is a highly radioactive metal about which little is known. Predict:
a) its density b) its melting point c) its boiling point

2 Write a balanced formula equation and ionic equation for the following reactions:
a) potassium and bromine b) lithium and water c) sodium and oxygen

3 Which is more reactive, rubidium or lithium? Explain in terms of electron loss.

4 Which is more reactive, sodium or magnesium (2.8.2)? Explain by electron loss.

5 Write the chemical formula and the ionic formula of:
a) rubidium sulphate b) sodium nitrate c) potassium iodide.

The first transition metals

- The first transition metals are a series of metallic elements lying between Groups II and III of the periodic table.
- Most are hard, heavy, industrially-important metals.

A Properties

1 Basic data for the first transition metals

name	symbol	atomic number	electronic configuration	density (g/cm^3)	melting point (°C)	boiling point (°C)
scandium	Sc	21	2.8.9.2	2.99	1540	2830
titanium	Ti	22	2.8.10.2	4.50	1670	3290
vanadium	V	23	2.8.11.2	5.96	1800	3380
chromium	Cr	24	2.8.12.2	7.20	1850	2260
manganese	Mn	25	2.8.13.2	7.40	1245	1900
iron	Fe	26	2.8.14.2	7.87	1530	2750
cobalt	Co	27	2.8.15.2	8.90	1490	2870
nickel	Ni	28	2.8.16.2	8.90	1450	2730
copper	Cu	29	2.8.17.2	8.94	1083	2630
zinc	Zn	30	2.8.18.2	7.15	420	910

2 Transition metals are **good conductors** of **heat** and **electricity** because the spaces in the incomplete penultimate shell allow electron movement.

- Although zinc is included in the series, its inner electron shells are complete and it therefore sometimes has different properties (lower melting point, boiling point, white compounds only; single valency, forming Zn^{2+} ions only).

3 They have **strong inter-atomic forces** because of the unfilled shell and so have **high melting and boiling points.**

4 They have **similar size atoms** because the number of electron shells in use is the same across the series.

5 Moving **left to right** there is a trend to **slightly smaller atoms** because the increasing number of protons tends to **attract the electrons in closer**. This leads to **increasing density left to right** as the smaller atoms pack closer.

6 Iron, cobalt, nickel are strongly magnetic.

Q Why do the transition metals have such high melting and boiling points?

KEY FACT

7 They are all metals because they all form positive ions by loss of the two outer electrons: $M - 2e^- \rightarrow M^{2+}$

B Compounds and reactions

KEY FACTS

1 Transition metals are not as reactive as the Group I and II metals because electron loss is not as easy as it is for the Group I and II metals.

2 This lower reactivity means they can be extracted from their ores by chemical reduction.

Remember
An alkali is a soluble base.

3 Their oxides and hydroxides are less strongly basic than those of the Group I and II metals and are insoluble in water, and therefore do not form alkalis.

KEY FACT

4 It is characteristic of the transition metals that, as well as the two outer ones, they can lose electrons from the penultimate shell and so they can exhibit <u>variable valency</u>.

Examples: copper forms Cu^{2+} and Cu^{+} ions; iron forms Fe^{2+} and Fe^{3+} ions; cobalt forms Co^{2+} and Co^{3+} ions.

KEY FACT

5 Unlike the Group I and II metals, which all have white compounds forming colourless solutions, the transition metals have <u>coloured compounds and solutions</u>.

Examples: Cu^{2+}(blue), Fe^{2+}(green), Mn^{2+}(pink), Cr^{3+}(blue)

6 Transition metals use the unfilled spaces in the penultimate shell to accept electron sharing from oxygen atoms and form complex negative oxo-ions.

Examples: chromate CrO_4^{2-}, manganate MnO_4^{-}, dichromate $Cr_2O_7^{2-}$

Q Why are the oxides of the transition metals not alkalis?

7 Transition metals have multiple uses.

Examples: titanium (aircraft parts, replacement hip joints), chromium (alloys, plating), copper (wiring, piping), zinc (protection by galvanizing), and iron (iron/steel structures).

1 Explain why the compounds of zinc are white, whilst the other transition metals have coloured compounds.

2 Explain why the density of the transition metals increases from left to right across the series.

3 Why do we regard the elements in the transition block as metals?

4 Why do the transition metals have similar-sized atoms across the series?

5 Why can the transition metals have variable valency?

Group VII – the halogens

THE BARE BONES

- The halogens make up Group VII of the periodic table, and include fluorine, chlorine, bromine and iodine.
- Their reactivity decreases as you go down the group.

A Properties and reactions

1 Basic data for halogens:

name	symbol	atomic number	electronic configuration	colour and physical state at room temp and pressure	melting point (°C)	boiling point (°C)
fluorine	F	9	2.7	pale yellow gas	-220	-188
chlorine	Cl	17	2.8.7	green/yellow gas	-101	-35
bromine	Br	35	2.8.18.7	dark red liquid	-7	59
iodine	I	53	2.8.18.18.7	shiny purple crystal	114	187
astatine	At	85	2.8.18.32.18.7	?	?	?
				darker ↓	increase ↓	increase ↓

KEY FACT

2 All are diatomic molecules X_2. The two atoms are joined by a covalent bond i.e.

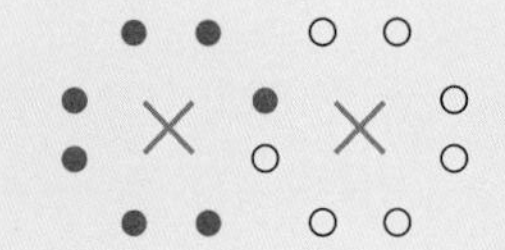

- Each molecule is separate from each other molecule and intermolecular forces are so low that they are easily overcome, leaving the molecules free to move randomly.
- So, all the halogens have extremely low melting and boiling points.

KEY FACT

- As atomic number increases down the group, the size and total mass of the X_2 molecule increases; intermolecular forces between these larger, heavier molecules increase; hence melting point and boiling point increase down the group.

3 The halogens react chemically with:

- the alkali metals to form ionic halides:

$$2M(s) + X_2(g) \rightarrow 2MX[2M^+X^-](s)$$

Example: $2Na(s) + Br_2(l) \rightarrow 2NaBr[2Na^+Br^-](s)$

- hydrogen to form covalent hydrogen halides:

$$H_2(g) + X_2(g) \rightarrow 2HX(g)$$

Example: $H_2(g) + Cl_2(g) \rightarrow 2HCl(g)$

Remember
Hydrogen halides ionise completely in aqueous solution and are strong acids.

A

- **reducing agents** in **redox reactions**; the **halogen is the oxidising agent** (see page 92).

Example:

OXIDATION: gain of oxygen (Br₂ → S)

$$H_2S(g) + Br_2(l) \rightarrow 2HBr(g) + S(s)$$

reducing agent (H_2S), oxidising agent (Br_2)

loss of hydrogen: OXIDATION (H_2S → S)

4 The hydrogen halides **ionise completely** in aqueous solution to provide **H^+ ions** and so are **strong acids**:

$$HCl(g) \xrightarrow{H_2O} HCl(aq) \rightarrow H^+(aq) + Cl^-(aq) \text{ (hydrochloric acid)}$$

Q What is a diatomic molecule?

B Comparative reactivity of the halogens

1 The **reactivity** of the halogens **decreases down the group**.

2 This is best demonstrated by **displacement** reactions in which **more reactive halogens displace less reactive halogens** from a solution of the ionic compound of the less reactive halogen.

3 Chlorine displaces bromine from bromides and iodine from iodides, whilst bromine displaces iodine from iodides. The simplest equations are:

$$Cl_2(g) + 2Br^-(aq) \rightarrow 2Cl^-(aq) + Br_2(l)$$
$$Cl_2(g) + 2I^-(aq) \rightarrow 2Cl^-(aq) + I_2(s)$$
$$Br_2(e) + 2I^-(aq) \rightarrow 2Br^-(aq) + I_2(s)$$

4 It is clear from the above that the **reactivity of the halogens is a measure of the ease of formation of the uninegative ion**:

$$X(atom) + e^- \rightarrow X^- \text{ (uninegative halide ion)}$$

5 Formation of this ion is **easiest** at the **top** of the group (fluorine) and **most difficult** at the **bottom** (iodine) because, at the top of the group, the electron is going to an orbit which is:

- near to the positive nucleus, thus increasing the force of attraction
- less shielded from the attracting nucleus by inner electron shells

Remember
These displacement reactions are also redox reactions i.e.
$Cl_2 + 2e^- \rightarrow 2Cl^-$ (reduction)
$2Br^- - 2e^- \rightarrow Br_2$ (oxidation).

Q Which is more reactive, chlorine or bromine?

1 Write the simplest ionic equation – or write 'no reaction' – for the following pairs of reactants:
a) chlorine and potassium bromide b) iodine and sodium chloride
c) bromine and potassium iodide

2 Explain why the reactivity of the halogens decreases with increasing atomic number and why this is the reverse of the situation for the alkali metals.

3 Which is more reactive, chlorine (2.8.7), or oxygen (2.8.6)? Explain your answer.

Group 0 and the periodic table

THE BARE BONES

- The noble gases make up Group 0 of the periodic table, and include helium, neon and argon.
- All are very unreactive gases.

A Properties

1 Basic data for noble gases

name	symbol	atomic number	electronic configuration	density (g/cm^3)	melting point (°C)	boiling point (°C)
helium	He	2	2	0.18	-271.4	-269
neon	Ne	10	2.8	0.90	-248.7	-246
argon	Ar	18	2.8.8	1.78	-189.2	-186
krypton	Kr	36	2.8.18.8	3.74	-157	-153
xenon	Xe	54	2.8.18.18.8	5.90	-112	-108
radon	Rn	86	2.8.18.32.18.8	9.96	-71	-62
				increase	increase	increase

2 The low melting and boiling points and densities increase down the group as the atom gets heavier and inter-atomic forces increase.

3 All are present in very small amounts in the atmosphere and can be obtained by the fractional distillation of liquefied air.

KEY FACTS

4 All are unreactive because their highest occupied energy level is full so the atoms have no tendency to lose, gain or share electrons.

5 This inertness means they are monatomic - exist in single atoms - and not diatomic like other gaseous elements.

6 The noble gases are used as follows:

- Helium - less dense than air and does not burn, so used in balloons and airships.
- Neon - glows brightly in electrical discharge, so used in advertising signs.
- Argon - in domestic light bulbs, as its inertness prevents the filament burning.

Q What is 'monatomic' and why are the noble gases monatomic?

PRACTICE

1 Explain why the melting point and boiling point of the noble gases increase with increasing atomic number.

2 Explain why the noble gases are inert and unreactive.

B The full periodic table

I [1]	II [2]											III	IV	V	VI	VII [3]	0 [4]
							H Hydrogen 1										He Helium 2
Li Lithium 3	Be Beryllium 4											B Boron 5	C Carbon 6	N Nitrogen 7	O Oxygen 8	F Fluorine 9	Ne Neon 10
Na Sodium 11	Mg Magnesium 12					transition metals						Al Aluminium 13	Si Silicon 14	P Phosphorus 15	S Sulphur 16	Cl Chlorine 17	Ar Argon 18
K Potassium 19	Ca Calcium 20	Sc Scandium 21	Ti Titanium 22	V Vanadium 23	Cr Chromium 24	Mn Manganese 25	Fe Iron 26	Co Cobalt 27	Ni Nickel 28	Cu Copper 29	Zn Zinc 30	Ga Gallium 31	Ge Germanium 32	As Arsenic 33	Se Selenium 34	Br Bromine 35	Kr Krypton 36
Rb Rubidium 37	Sr Strontium 38	Y Yttrium 39	Zr Zirconium 40	Nb Niobium 41	Mo Molybdenum 42	Tc Technetium 43	Ru Ruthenium 44	Rh Rhodium 45	Pd Palladium 46	Ag Silver 47	Cd Cadmium 48	In Indium 49	Sn Tin 50	Sb Antimony 51	Te Tellurium 52	I Iodine 53	Xe Xenon 54
Cs Caesium 55	Ba Barium 56	La Lanthanum 57 x	Hf Hafnium 72	Ta Tantalum 73	W Tungsten 74	Re Rhenium 75	Os Osmium 76	Ir Iridium 77	Pt Platinum 78	Au Gold 79	Hg Mercury 80	Tl Thallium 81	Pb Lead 82	Bi Bismuth 83	Po Polonium 84	At Astatine 85	Rn Radon 86
Fr Francium 87	Ra Radium 88	Ac Actinium 89 •															

This line divides the metals from the non-metals.

x Lanthanide series	Ce Cerium 58	Pr Praseodymium 59	Nd Neodymium 60	Pm Promethium 61	Sm Samarium 62	Eu Europium 63	Gd Gadolinium 64	Tb Terbium 65	Dy Dysprosium 66	Ho Holmium 67	Er Erbium 68	Tm Thulium 69	Yb Ytterbium 70	Lu Lutetium 71
• Actinide series	Th Thorium 90	Pa Protactinium 91	U Uranium 92	Np Neptunium 93	Pu Plutonium 94	Am Americium 95	Cm Curium 96	Bk Berkelium 97	Cf Californium 98	Es Einsteinium 99	Fm Fermium 100	Md Mendelevium 101	No Nobelium 102	Lr Lawrencium 103

KEY

X element name z

X = atomic symbol
z = atomic number

[1] Group I comprises the alkali metals. [2] Group II comprises the alkaline-earth metals. [3] Group VII comprises the halogens. [4] Group 0 comprises the noble gases.

Rates of reaction

THE BARE BONES

- Reactions can only occur when particles collide with enough energy, called the activation energy.
- Factors such as temperature, concentration and catalysts affect the speed of chemical reactions.

A Reaction rate and temperature

1 Reactions can only occur when particles collide with enough energy.

KEY FACT

2 This energy is called the activation energy (E_{Act}) of the reaction.

- Collisions which have E_{Act} and thus lead to reaction are called **fruitful collisions**.

Q How does a temperature increase affect particles?

3 **Increasing the temperature** of the reactants increases the **speed** of the reactant particles.

- The particles **collide more often** and with **higher energy**.
- This **increases** the number of **fruitful collisions**.

KEY FACT

4 Hence, rate increases as temperature increases.

B Concentration and rate

1 **Increasing the concentration** of reactants in solution **increases the number** of **reactant particles** available.

2 This **increases the frequency** of **fruitful collisions** and, hence, **rate**.

KEY FACT

3 Rate increases as concentration increases.

4 **Example:** $Zn(s) + 2HCl(aq) \rightarrow ZnCl_2(aq) + H_2(g)$

- The more concentrated the acid, the higher the reaction rate.

Q Increasing pressure on gases causes concentration increase. Why?

5 For reactions involving **gases**, **increasing the pressure** is effectively **increasing concentration** as the particles are **closer** and **fruitful collisions more frequent**.

C Surface area and rate

Q On adding hydrochloric acid to powdered marble, it fizzes out of the test tube. Why?

1 When marble chips are reacted with hydrochloric acid:

$$CaCO_3(s) + 2HCl(aq) \rightarrow CaCl_2(aq) + H_2O(l) + CO_2(g)$$

- The **smaller** the pieces of marble used, the **faster** the reaction.

2 When the solid is **powdered**, **more surface** and **more particles** are **exposed to the acid**, thus **increasing** the number of **fruitful collisions**.

D Catalysts

KEY FACT

1 A catalyst increases the rate of a reaction but is not itself used up during the reaction. It can be used over and over again.

2 Different reactions need different catalysts.

Examples: $2H_2O_2(l) \xrightarrow[\text{catalyst}]{MnO_2} 2H_2O(l) + O_2(g)$
hydrogen peroxide

$N_2(g) + 3H_2(g) \xrightarrow[\text{catalyst}]{Fe} 2NH_3(g)$

KEY FACT

3 Catalysts do not increase the number of fruitful collisions but work by reducing the activation energy required, i.e.:

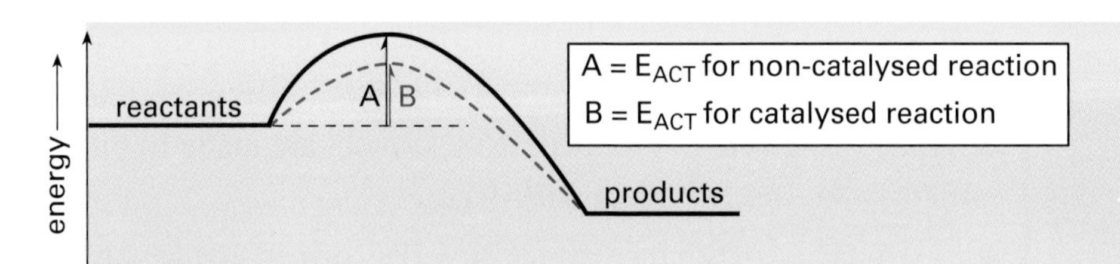

Q Give a definition of a 'catalyst'.

E Enzymes

KEY FACT

1 Living things produce enzymes which are proteins that catalyse the body's reactions.

2 A cell may contain a thousand enzymes and each one is specific and catalyses only one reaction.

3 Enzymes work by fitting like jigsaw pieces into the reactant molecule. They will not fit into other molecules.

enzyme + reactant → intermediate product —breaks up→ enzyme released unchanged (a catalyst) + products

4 Enzymes are destroyed above 45°C and work best at about 40°C and in a pH of about 6.5.

5 They are widely used in industry for fermentation, production of yoghurt, cheese making, vitamin production, biological washing powders.

Q What is an enzyme?

PRACTICE

1 Use the collision theory to explain the requirement for a reaction to occur.

2 Explain the effect of a) temperature b) surface area, on rates of reaction.

3 Catalysts used to slow down chemical reactions are called 'inhibitors'. Explain how you think they might work.

Energy in reactions

- Energy is released when a chemical bond is formed; conversely, to break a bond, energy must be supplied.
- Chemical reactions involve the making and breaking of bonds.
- The heat given out or taken in during a chemical reaction can be calculated by knowing which bonds are involved.

A Bond energies

KEY FACT

1 The temperature changes observed in chemical reactions result from the making and breaking of bonds.

2 Reactions begin by breaking the bonds between the atoms of the reactant molecules. This requires energy to be put in.

3 When products are formed, new bonds are made to build the product molecules. This releases energy.

KEY FACT

4 In an exothermic reaction, the energy released from forming new bonds is greater than the energy required to break bonds in the reactant molecules.

Remember
break bonds - energy in;
make bonds - energy out.

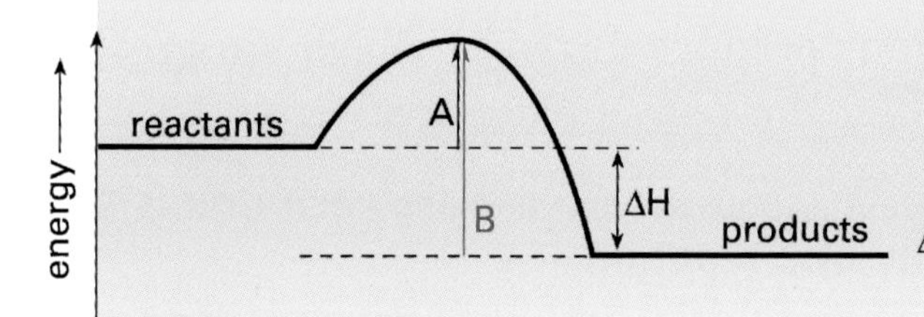

A = Activation energy – heat required to break bonds
B = energy released from forming bonds
ΔH = heat given out in the reaction

The energy content of the reactants is greater than the energy content of the products in an exothermic reaction e.g. combustion, neutralisation.

KEY FACT

5 In an endothermic reaction, the energy released from forming new bonds is less than the energy required to break bonds in the reactants.

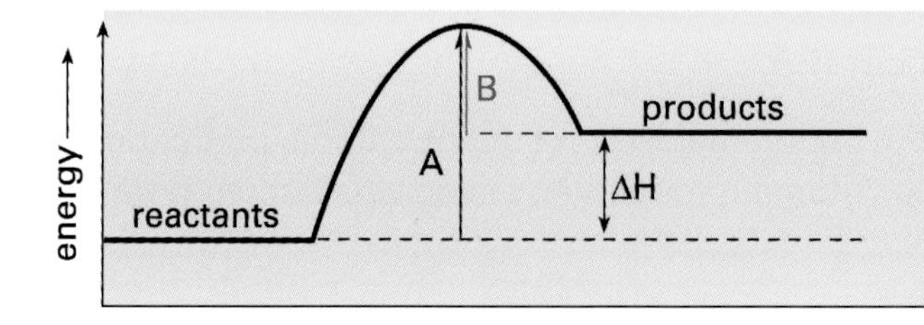

The energy content of the reactants is less than that of the products e.g. photosynthesis, thermal decomposition.

These diagrams are energy level diagrams, and you will need to know how to interpret them.

Q Define 'exothermic' and 'endothermic'.

B Heat of reaction (ΔH)

KEY FACTS

1 A change of energy content is given the symbol ΔH (measured in kJ/mole).

2 In a reaction, the change in heat (ΔH) is determined by:
ΔH = energy of products − energy of reactants.

3 In an exothermic reaction, where energy of products is less than energy of reactants, ΔH is negative.

4 In an endothermic reaction, where energy of products is greater than energy of reactants, ΔH is positive.

Q What does the symbol ΔH represent?

C Bond energy values

KEY FACTS

1 The energy of a bond is the amount of energy (+kJ/mol) required to break the bond.

2 The same amount of energy is released when a bond is formed.

3 Some bond energies (in kJ/mol) are: H−O (460); C=C (610); O=O (500); H−H (435); C−C (350); C−H (410); C−O (360).

4 Example: Calculate the heat of reaction for the combustion of hydrogen.

- The balanced equation is $2H_2(g) + O_2(g) \rightarrow 2H_2O(g)$.
- Bonds to be **broken** are:

2 H−H	=	2 × 435	=	+870 kJ/mol (using values above)
1 O=O		1 × 500	=	+500 kJ/mol
total energy input			=	+1370 kJ/mol.

- Bonds to be **formed** are: 4 O−H = 4 × (−460)
  Total energy given out = −1840 kJ/mol.
- Total ΔH for reaction = ΔH breaking + ΔH forming
  = +1370 + (−1840)
  = −470 kJ/mol.
- the heat of reaction is therefore **−470 kJ/mol** and the reaction is **exothermic**.

Remember
Bond energy values are negative when a bond is formed.

Q Give a definition of 'bond energy'

PRACTICE

1 The equation for the complete combustion of methane is:

$CH_4(g) + 2O_2(g) \rightarrow CO_2(g) + 2H_2O(g)$ ΔH is negative

Which has the greater heat content, the reactants or the products?

Reversible reactions

THE BARE BONES

- ➤ Reversible reactions take place in both directions simultaneously.
- ➤ Reversible reactions proceed to a point of equilibrium.
- ➤ Many industrial processes involve reversible reactions.

A Basic principles

KEY FACT

1 Reversible reactions can be represented as: A + B ⇔ C + D.

2 Reversible reactions are when the products can react together to give the reactants.

3 In a reversible reaction, the forward reaction (A + B → C + D) and backward reaction (A + B ← C + D) take place simultaneously.

Q Define the term 'reversible reaction'.

B Equilibrium

KEY FACTS

1 When the rate of the forward reaction equals the rate of the backward reaction there is no further change in the amounts of reactants and products present. At this point, the reaction has reached equilibrium.

2 It may be that equilibrium is reached at, say, 30% conversion to C and D or, under different conditions, at say 60% conversion to C and D.

3 This is described as changing the position of the equilibrium.

Q What is a 'reversible reaction' and when does it reach equilibrium?

going up at same rate as escalator coming down – equilibrium reached 30% of way up

going up at same rate as escalator coming down – equilibrium reached – 60% of way up

The position of the equilibrium has changed

KEY FACTS

4 Predicting how the position of equilibrium of a reversible reaction will change in response to changing the conditions is governed by Le Chatelier's Principle.

5 This states that when conditions are changed in a system at equilibrium, the equilibrium will move in such a way as to minimise the effects of the change.

C The Haber process

Ammonia is manufactured from nitrogen and hydrogen (see page 102).

1 The reaction is exothermic.

$$N_2(g) + 3H_2(g) \rightarrow 2NH_3(g): \Delta H \text{ is negative}$$

2 Unfortunately, ammonia is easily decomposed into nitrogen and hydrogen.

The reaction is now endothermic by exactly the same amount.

$$2NH_3(g) \rightarrow N_2(g) + 3H_2(g): \Delta H \text{ is positive}$$

3 The reaction is therefore reversible and the complete equation is:

$$N_2(g) + 3H_2(g) \rightleftharpoons 2NH_3(g) + \text{heat } (\Delta H \text{ is negative; exothermic})$$

4 The effect of temperature on this equilibrium is as follows:

- an increase in temperature will, according to Le Chatelier, result in the system trying to absorb the heat - it therefore moves right to left where the reaction is endothermic - so the yield of ammonia decreases.
- increasing temperature, however, increases the number of fruitful collisions and therefore increases rate (see page 118).
- high temperature therefore means low yield but high rate.

5 According to Le Chatelier, if pressure increases, the system will minimise this effect by moving to fewer molecules and lower volume – i.e. move to the right.

- Increasing pressure also pushes the molecules closer together, thus increasing the number of fruitful collisions.
- High pressure therefore causes high yield and high rate of reaction.

6 A catalyst increases rate, but there is no effect on the equilibrium.

7 The conditions to give the best compromise of yield, rate and cost are:

$$N_2(g) + 3H_2(g) \xrightleftharpoons[\text{500 – 550°C, Fe catalyst}]{\text{250 atmospheres}} 2NH_3(g) \quad \Delta H \text{ is negative}$$

8 This yields about 15% conversion to ammonia.

Q State Le Chatelier's principle.

PRACTICE

1 The reaction ($A(g) + 2B(g) \rightleftharpoons C(g) + D(g)$: ΔH is negative) is left to reach equilibrium in a sealed container. Explaining your answers, what will be the effect on the position of the equilibrium of:

a) increasing temperature
b) increasing pressure
c) inserting a catalyst
d) removing product D
e) removing reactant B?

Current, voltage and resistance

- ➤ Electric current transfers energy round a circuit.
- ➤ An electric current is a flow of charge and is measured in amperes (A).
- ➤ Voltage measures the difference in the energy carried by the charge between the two points.
- ➤ Resistance measures how difficult it is for a current to pass round the circuit.

A Electric current and charge

KEY FACT

1 Electric current is a flow of charged particles around an electric circuit.

2 The charges transfer energy from a battery or other power supply to components in the circuit, such as lamps and resistors.

3 In wires and most other solids, electric current is carried by negatively charged electrons. In electrolytes (e.g. salt dissolved in water), current is carried by positive and negative ions.

4 Electric current I is measured in amperes (A). Electric current is measured with an ammeter, which is placed in the circuit in series.

5 Electric charge Q is measured in coulombs (C). When the current is 1 ampere a charge of 1 coulomb passes in a second.

charge (coulombs) = current (amperes) × time (seconds)

$$Q = It$$

Remember
You need to be able to recall and use Q = It.

Q Calculate the current that passes through a cell when 180 C flows in 6 minutes.

B Voltage

1 A current only passes through a component or wire if there is a voltage across it.

KEY FACT

The bigger the voltage across a component, the bigger the current through that component.

2 Voltage V is measured in volts (V).

Voltage is measured with a voltmeter, which is placed in parallel with the component.

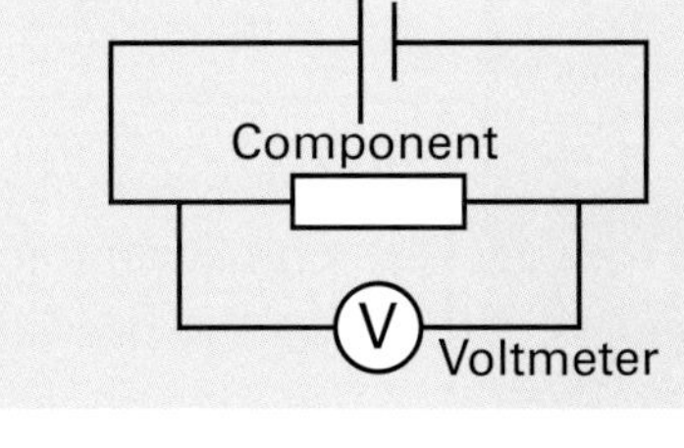

3 Voltage is also called potential difference. It measures the amount of energy transferred per unit charge that flows. 1 volt = 1 joule per coulomb.

energy transferred (joules) = voltage (volts) × charge (coulombs)

$$E = VQ$$

Remember
You need to be able to recall and use E = VQ.

Q Can you identify all the components in the circuits on this page?

C Resistance

1 Resistance in a wire or component makes it more difficult for a current to pass.

KEY FACT

2 Changing the resistance in a circuit changes the current. The bigger the resistance in the circuit, the smaller the current (if the voltage does not change).

3 Variable resistors can be used to control the current in a circuit. The longer the wire, the greater the resistance, and the smaller the current. A lamp in the circuit will glow less brightly.

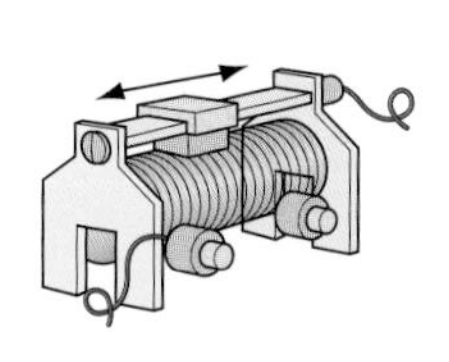

Variable resistor

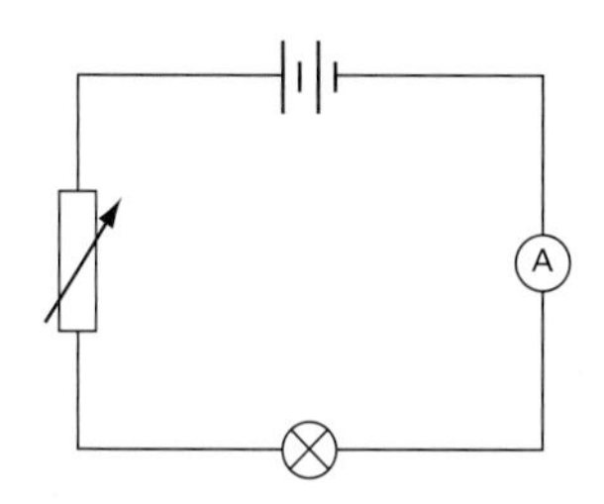

Circuit with variable resistor, lamp and ammeter.

4 Resistors get hot when a current passes through them. The filament in a lamp is a very thin wire, which gets hot and glows when a current passes. The same effect heats the element in an electric fire or a hairdryer.

Resistance is measured in ohms (Ω). An important equation connects voltage, current and resistance, which can be written in more than one way:

$$\text{voltage (volts)} = \text{current (amps)} \times \text{resistance (ohms)}$$

$$V = IR$$

or,

$$\text{resistance (ohms)} = \frac{\text{voltage (volts)}}{\text{current (amperes)}}$$

$$R = \frac{V}{I}$$

Example: The voltage across a lamp filament the lamp is 12 V when the current is 3 A. Calculate the resistance of the filament.

Answer: $R = \frac{V}{I} = \frac{12\text{ V}}{3\text{ A}} = 4\,\Omega$

Remember to show your working when you are doing calculations.

Remember
You need to be able to recall and use V = IR.

Q What is the voltage across a 10 Ω resistor when the current is 2 A?

PRACTICE

1 A current of 4 A passes through a motor. The motor runs for 5 minutes.
a) Calculate the charge that passes through the motor in 5 minutes.
b) The voltage across the motor is 24 V. Calculate the energy transferred to the motor in 5 minutes.

2 Calculate the voltage across a 5 Ω resistor when the current passing through it is 2 A.

Electric components 1

- The electric current in a circuit is controlled by the voltage of the supply and the resistance of the components in the circuit.
- Current–voltage (I–V) graphs are used to show how the current through a component depends on the voltage across it.
- Different components have differently-shaped current–voltage graphs.

A Resistors

KEY FACT

1 A resistor is a component that is designed to have a constant resistance.

2 The **electric current through a resistor** (and other circuit components) **depends on the voltage across it.**

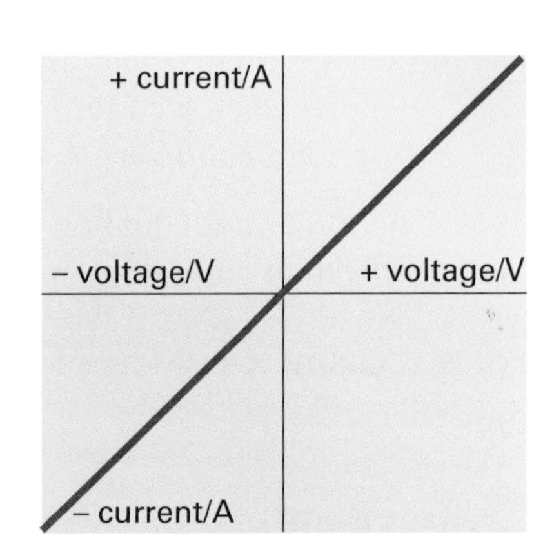

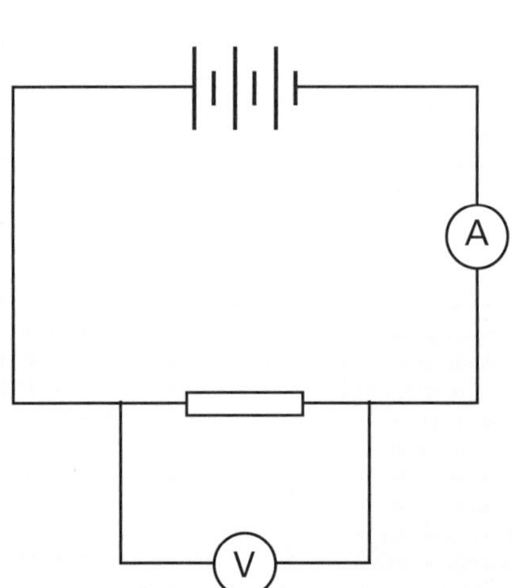

Resistance is $\frac{\text{voltage}}{\text{current}}$. The graph is a straight line so $\frac{V}{I}$ is constant.

Q What is the equation linking V, I and R?

Always check the axes of a graph – which quantity is being plotted on which axis?

B Lamps

1 The resistance of metals increases as the temperature increases.

2 In a **lamp**, the **filament gets hotter** as the **voltage across it increases**. As it gets hotter, its atoms vibrate more. It is **more difficult for the current to flow** and so its **resistance increases.**

As the voltage increases, the current does not increase as much – because the resistance has increased.

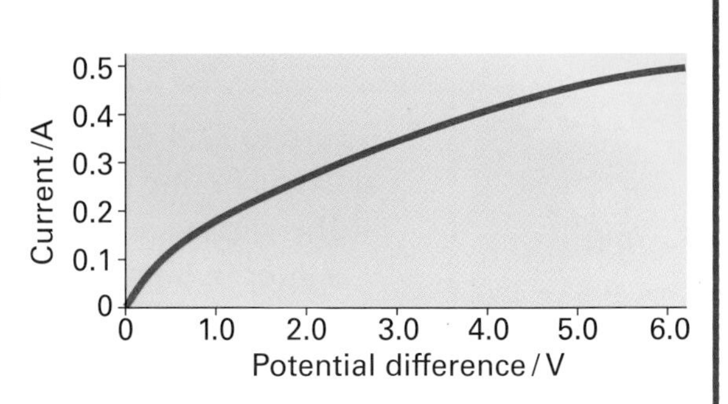

B

3 Example: Use the current–voltage graph on page 127 to calculate the resistance of the lamp when the voltage is 2 V and when it is 4 V. Explain why the resistance values are different.

Answer: Use the graph to find the current values for V = 2 V and V = 4 V. Calculate the values of resistance:

$$R = \frac{V}{I} = \frac{2\ \text{V}}{0.27\ \text{A}} = 7.4\ \Omega \qquad R = \frac{V}{I} = \frac{4\ \text{V}}{0.42\ \text{A}} = 9.5\ \Omega$$

The wire becomes warm as the current passes through. As it becomes warm the particles in the metal vibrate more, making it more difficult for the conduction electrons to pass through.

Q How does the resistance of a wire change as it gets hot?

C Diodes

KEY FACT

A diode only lets current pass one way. When the voltage is reversed no current passes.

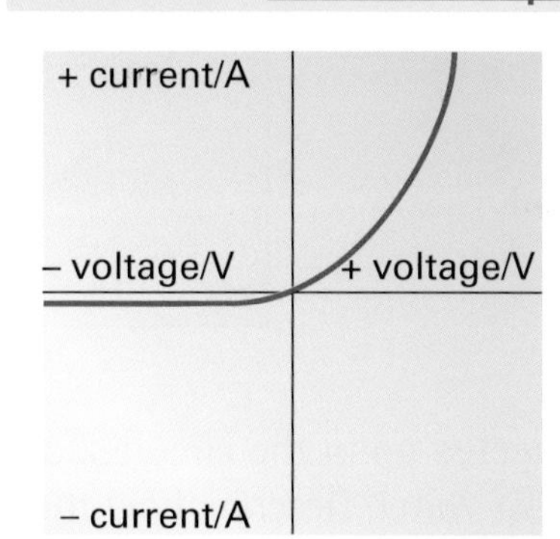

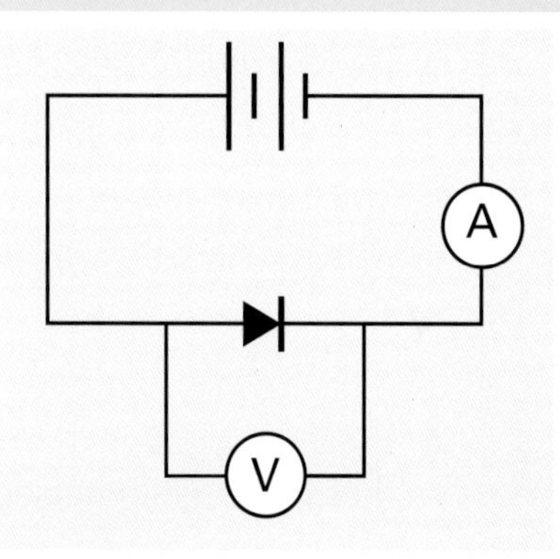

No current flows when the voltage is reversed.

Q Why is a diode sometimes called a valve?

PRACTICE

1 Copy the current-voltage graph for the resistor and draw a second line to show how the current changes with voltage across a resistor with **twice** the resistance.

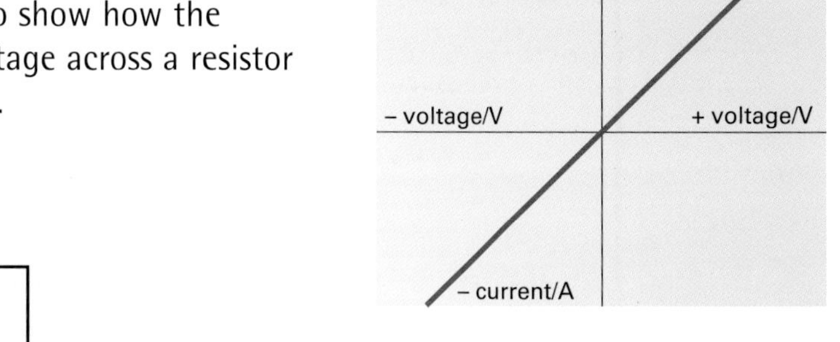

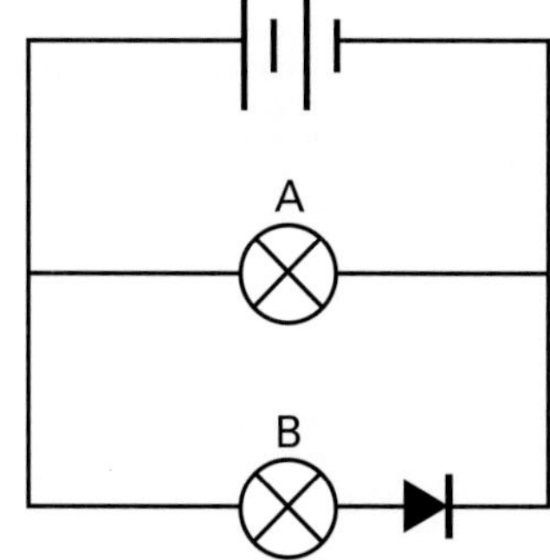

2 The diode in the circuit on the left allows the current to pass through it. What will happen to lamps A and B if the diode is reversed?

Electric components 2

THE BARE BONES

- ➤ The resistance of a thermistor decreases as it gets warmer.
- ➤ The resistance of a light dependent resistor (LDR) decreases as the light intensity increases.

A Thermistors

1 A thermistor is a semiconductor component that can be used to detect changes in temperature.

KEY FACT

- The resistance of a thermistor <u>gets less</u> as its <u>temperature increases</u>.

- The resistance of a semiconductor, such as silicon, decreases as it warms up.
- The energy ionises some of the atoms, releasing more electrons.
- More electrons flowing in the circuit, which means the current increases.

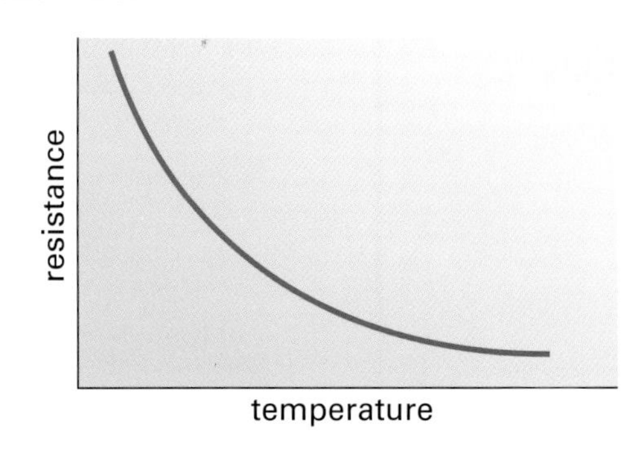

2 **Example:** A thermistor is connected up in the circuit shown below. The thermistor is immersed in a beaker of water. Describe how the current in the circuit will change as the water is warmed up.

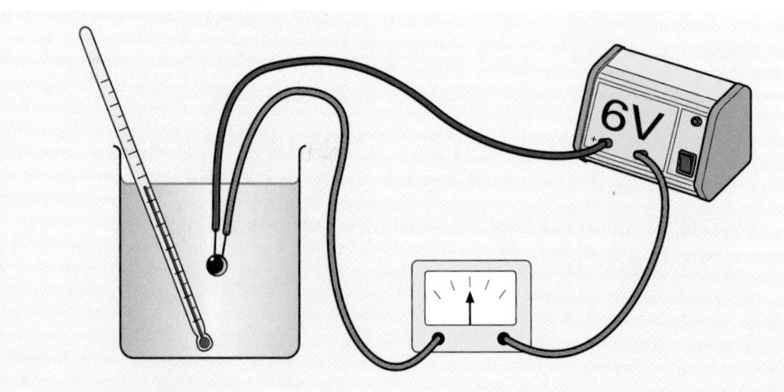

Answer: The resistance-temperature graph shows that the resistance of the thermistor decreases as the temperature rises. If the resistance in the circuit decreases, the current will increase.

Q Can you draw the circuit diagram for the thermistor circuit shown here?

B Light-dependent resistors (LDRs)

1 An LDR (light-dependent resistor) can be used to detect changes in light intensity.

KEY FACT

- The resistance of an LDR gets less as the light level increases.

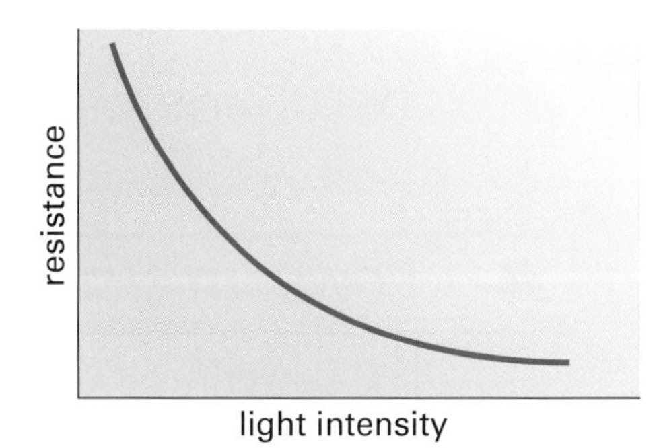

- The LDR is made of a semiconductor that has a very high resistance in the dark, but a much lower resistance when light shines on it.
- Energy from the light ionises atoms in the semiconductor, releasing more electrons.
- More electrons flowing in the circuit, which means the current increases.

Remember to label the axes on a graph – with quantities and units.

2 Example: An LDR is connected up in the circuit shown below. The LDR is moved towards the lamp. Describe how the current in the circuit will change, as the LDR gets closer to the lamp.

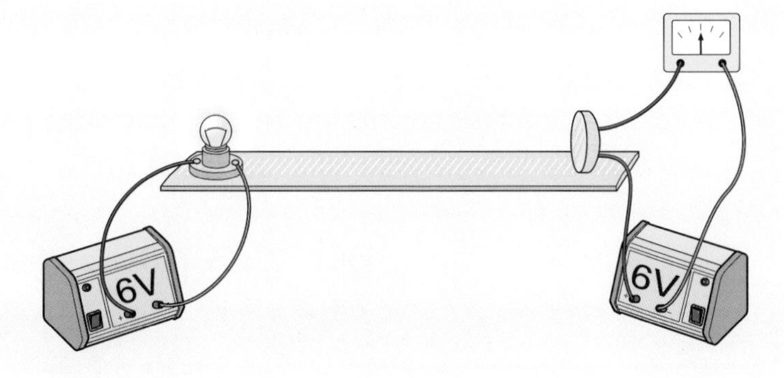

Answer: The resistance–light intensity graph shows that the resistance of the LDR decreases as the light intensity rises. If the resistance in the circuit decreases, the current will increase.

Q Draw and label the symbols for a resistor, a diode, an LDR and a thermistor.

PRACTICE

1 A thermistor is used to monitor the temperature in a greenhouse. Sketch a graph to show how the resistance of the thermistor might vary on a warm summer day.

2 Draw the circuit diagram for the LDR circuit shown in part 2 of Section B, above.

Electricity, energy and power

THE BARE BONES

- Electric current transfers energy from an energy source to the circuit.
- Power is the rate at which energy is transferred.
- The amount of energy transferred to an electrical appliance depends on how long it is switched on and the power of the appliance.
- Efficiency tells us how much of the power transferred in a process is useful.

A Power in electric circuits

1 Power is the rate at which energy is transferred. It is measured in watts (W) and kilowatts (kW).
1 kW = 1000 W.

- For more on power see Work, power and energy, page 168.

KEY FACT

2 In electric circuits,
power (watts) = voltage (volts) × current (amperes), or $P = VI$

3 **Example:** Philip uses an immersion heater to heat some water. The voltmeter reads 12 V and the ammeter reads 3 A. What is the power transferred from the power supply to the heater?

Answer:

power (W) = voltage (V) × current (A)
power = 12 V × 3 A = 36 W

Write down the equation you are going to use before you begin the calculation.

Q What is the power rating of a 3000 W electric heater in kW?

B Energy in electric circuits

Remember
You need to be able to recall and use $P = VI$.

1 Batteries, solar cells and generators are all sources of electrical energy. Electric current transfers the energy from the source to the circuit.

2 The amount of energy transferred to an electrical appliance depends on the time it is switched on and the power rating of the appliance.

- Energy transferred (joules) = power (watts) × time (seconds)
- Since power = voltage × current, you can also write:
  energy = voltage × current × time, or: $E = VIt$

3 **Example:** A 100 W immersion heater is used for 5 minutes. How much energy does it transfer?

Answer: 60 seconds in 1 minute so 5 minutes = 300 seconds
energy transferred (joules) = power (watts) × time (seconds)
energy transferred = 36 W × 300 s = 10 800 J

Q How much energy is transferred by a 60 W lamp in 10 minutes?

C Electrical energy in the home

1 Your home's **electricity meter** measures the **energy supplied** by the electricity company.

2 The meter measures energy in units called **kilowatt-hours** (kWh).

- One kilowatt-hour is the energy transferred when **1000 watts is transferred for one hour.**

KEY FACT

3 **The energy transferred in an electrical appliance with a given power rating can be calculated by:**
energy (kWh) = power (kW) × time (hours)

Remember
You need to be able to calculate the cost of running an electrical appliance.

4 **Example:** Electricity costs 6 pence per unit (1 kWh). What is the cost of running a 3 kW electric heater that is switched on for 4 hours?

Answer:
energy transferred in 4 hours
= 3 kW × 4 hours
= 12 kWh
cost = 12 kWh × 6 pence per kWh
= 72 pence

POWERGEN

Date of bill 31 August 01
Your customer reference number 012345678910
Quarterly Electricity Bill

Amount Due £125.34
Please pay now using the payment slip below

This is not a tax invoice

Charges for electricity used

	Present reading	Previous reading	Units used	Pence per unit	Charge amount
Domestic Economy 7					
Night	22236	21822	414	2.45	£10.14
Day	64443	62816	1627	5.94	£96.64
Standing charge from 1 Jun 01 to 30 Aug 01					£12.60
VAT @ 5.0%					£5.96
Amount Due					£125.34

Customer Service
Call free 0800 056 3856
If you have any questions simply phone us, we're open 24 hours a day, 7 days a week

Save £10 a year by changing to fixed month Direct Debit
If you complete and return the instructions on the reverse now you need not pay this bill and your monthly payment will be £62.00

Q How many units does a 2 kW kettle use in 6 minutes?

D Efficiency

1 Efficiency tells us how much of the power input becomes useful output.

2 $$\text{efficiency} = \frac{\text{useful power output}}{\text{power input}} \times 100\%$$

3 **Example:** A 40 W filament lamp gets very hot. In fact, 90% of the power transferred by the lamp warms up the room rather than lighting it. How much power is transferred to lighting the room?

Answer: 10% of the power from the lamp is useful to light the room.
10% of 40W is 4 W.

You need to be able to use the equation for efficiency.

Q What is the efficiency of a 3 kW motor that produces 1.5 kW of useful power?

PRACTICE

1 A current of 3 A passes through a 12 V lamp.
Calculate the energy transferred from the supply in 10 minutes.

2 A 15 W lamp transfers 6 W of its energy to light. Calculate the efficiency of the lamp.

3 A motor operates from a 12 V supply, with a current of 3 A. It is 75% efficient.
Calculate the useful energy transferred in 5 minutes.

Electricity at home

- Current can be either direct current or alternating current.
- Electric current supplies energy to the house through the live wire and it returns through the neutral wire, with a third earth wire for safety.
- Fuses and circuit breakers are designed to break a circuit if the current is too big.

A Direct current and alternating current

1 Electricity used in the home can be either **direct current** or **alternating current**.

2 **Batteries** supply a **direct current** (d.c.).

- Direct current is always in the **same direction**, and **does not change in size**.

KEY FACT

3 **Mains electricity supplies an alternating current (a.c.).**

- Alternating current is **constantly changing size and direction**.

4 **Example:** What is the frequency of the alternating current shown on the graph below?

The direction changes 50 times per second - the frequency is 50 Hz.

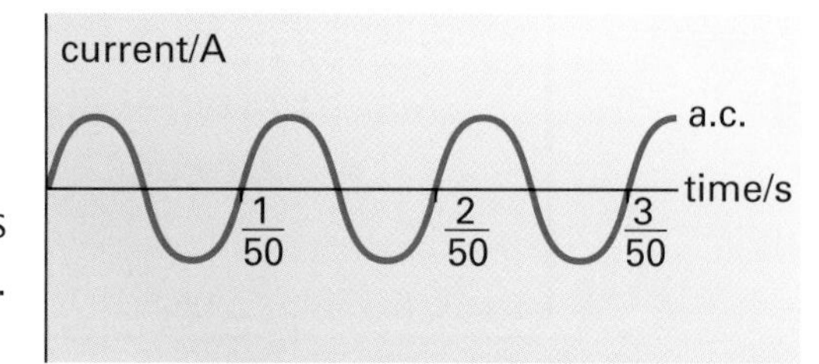

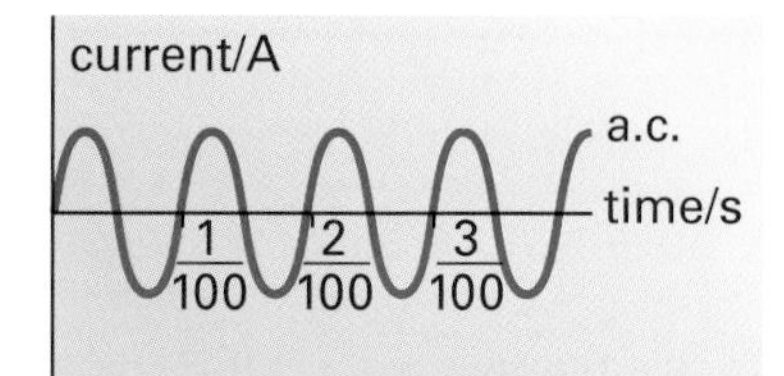

The current changes direction 100 times per second, so the frequency is 100 Hz.

Q What is the unit of frequency?

B Power cables and safety

Electricity is dangerous unless safety precautions are taken. For example, a small electric current passing across the heart is enough to kill! For safety, **power cables** have **three strands – live, neutral** and **earth**.

1 The electric current supplies energy to the house through the live wire and it returns through the neutral wire.

2 The live terminal alternates between positive and negative voltage, with the neutral terminal staying close to zero.

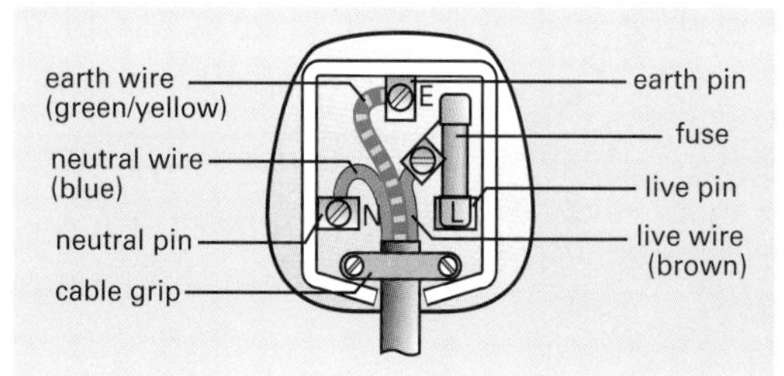

3 In normal use, the earth wire does not carry a current.

4 The strands are insulated to make sure that no current passes between the live and the neutral and earth wires.

Q What are the standard colours for the different wires in a plug?

C Fuses and circuit breakers

KEY FACT

1 Fuses and circuit breakers are safety devices designed to break the circuit if the current is too high.

- A fuse or circuit breaker near the electricity meter in your house **helps stop wiring overheating** and therefore helps to **prevent fire**.

2 The **fuse** in a **plug** is a thin resistance wire.

- The fuse melts if the current gets too large.
- The circuit breaks and prevents damage to the appliance.
- The fuse must match the current rating of the appliance.

3 Circuit breakers give **better protection than fuses**.

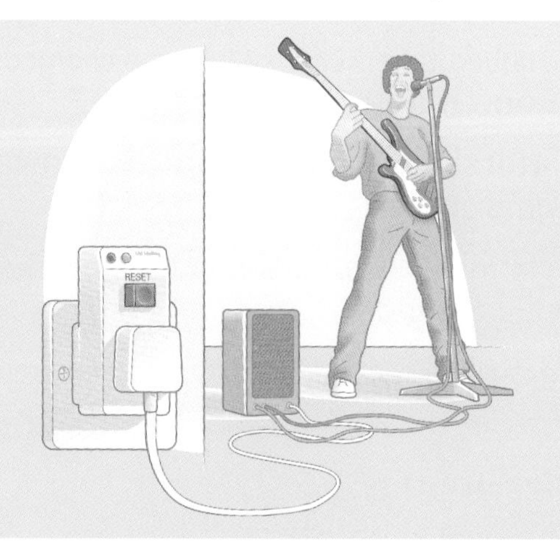

- The metal strings of the guitar are connected to the earth pin on the plug.
- If there is a fault and the strings become live, a large current might pass through the guitarist to reach earth.
- The large current will trip the circuit breaker.
- A circuit breaker will cut the circuit more quickly than a fuse.

4 Some appliances, such as hairdryers, are marked as having **double insulation** (see symbol on the right). There are no electrical connections to the casing of the appliance and no earth connection.

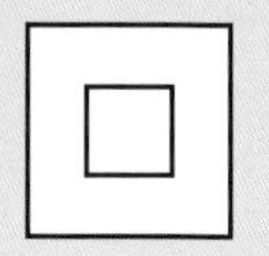

Electrical safety features are designed to stop wires overheating and currents passing through people.

Remember
You should be able to explain how a fuse and circuit breaker protect the circuit.

Q Why should the fuse rating be as close to the appliance current rating as possible?

PRACTICE

1 Suggest, with reasons, suitable materials for making each of the following parts of a cable and plug:
a) inner conducting core of cable
b) outer covering of cable
c) fuse in plug
d) casing of plug

2 Calculate the current that normally passes through a kettle rated 2 kW and connected to a 230 V supply. Should the plug have a 3 A fuse or a 13 A fuse? Explain your answer.

3 Describe the difference between direct current and alternating current.

Electric charge

THE BARE BONES

- ➤ Electric charge can be either positive or negative.
- ➤ Like charges repel each other. Opposite charges attract each other.
- ➤ Static electric charge is used in equipment, such as photocopying machines.

A Charge and electrostatics

1 Electric charge is ultimately a property of the electrons and protons that make up atoms. (*For more on electrons and protons see Atomic structure on page 74.*) Electrons are negatively charged while protons are positively charged.

2 When some materials are rubbed together they become charged. Electrons have been transferred from one material to the other.

- If the materials are insulators the charge does not leak away. The result is sometimes called static charge.
- The object that loses electrons will be positively charged. The object that gains electrons will be negatively charged.

3 The study of electric charges when they are not flowing in circuits is called electrostatics.

KEY FACT

Like charges repel each other.
Opposite charges attract each other.

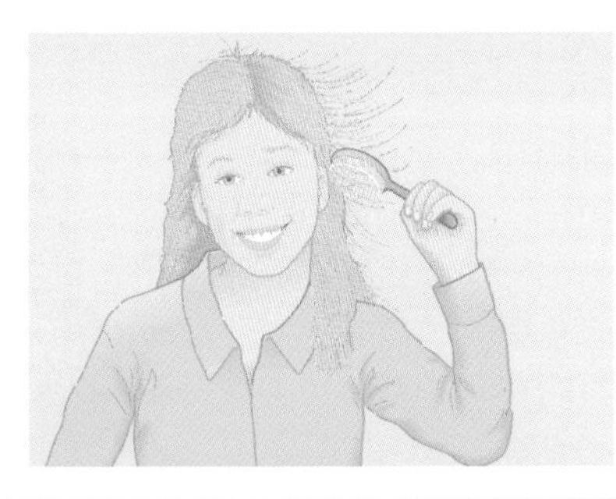

Charging your hair with a hairbrush makes it stand on end.

KEY FACT

4 When a charged object discharges it may do so with a spark – which can be dangerous.

- If a big enough charge builds up there will be a large voltage between the charged objects. Air is an insulator, but if the voltage is big enough the air conducts and a spark may be seen.
- Petrol passing along a refuelling pipe at a petrol station causes enough friction to give the pipe an electrostatic charge. If the charge builds up there is the danger of sparking and risk of an explosion. The pipe is connected to earth, so that the charge can leak away safely.

Q Can you explain why your hair stands on end and 'crackles'?

B Using electric charge

How a photocopier works

A photocopier and a laser printer both use electrostatic charge to print on paper.

- The page to be copied is placed face down on a sheet of glass.
- The surface of the drum is coated in a material that emits electrons when light falls on it.

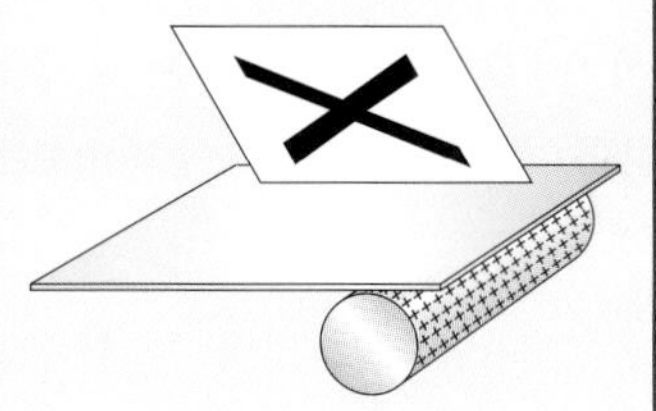

Remember
Like charges repel. Opposite charges attract.

- When the copier is switched on the surface of the drum becomes positively charged.
- A bright beam of light moves across the page. Light is reflected from white areas of the paper and reflected onto the drum below.
- Wherever light hits, electrons are emitted from the material of the drum and neutralise the positive charges. Dark areas on the page do not reflect light onto the drum, leaving a pattern of positive charges on the drum's surface.

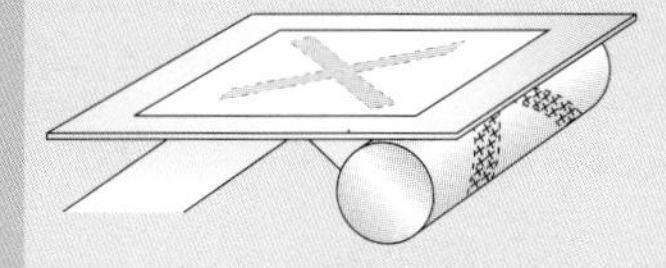

- Negatively charged, dry, black powder called toner is dusted over the surface of the drum, and the pigment particles are attracted to the positive charges.

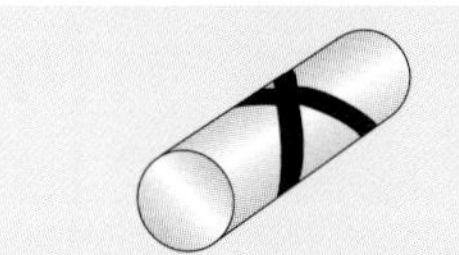

- A positively-charged sheet of paper then passes over the surface of the drum, attracting the beads of toner away from it.
- The paper is then heated and pressed to bond the image formed by the toner to the paper's surface.

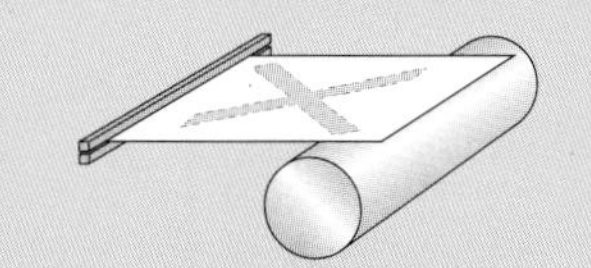

Removing ash from power station smoke

Electric charge is used to remove ash from the smoke in a coal-fired power station, using long metal electrodes fitted inside the chimney. The ash particles in the smoke, which have positive and negative ions attached, are attracted to the electrodes and removed from the smoke.

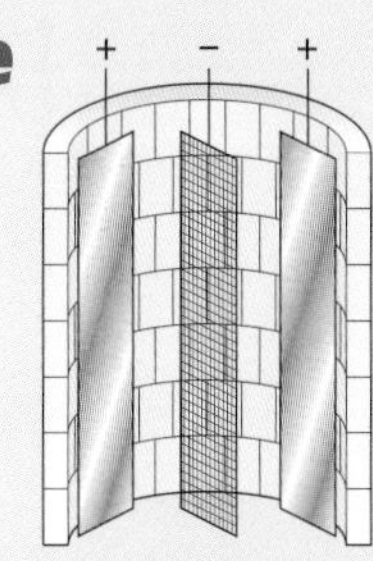

Q Can you explain why TV screens get very dusty quickly?

In current-charge calculations remember to use the time in seconds.

PRACTICE

1. When you take off a jumper in the dark you might see and hear sparks. Explain why.
2. Julie walks across a wool carpet and then touches a metal rail. She feels a small electric shock. Explain why.
3. Whilst an aircraft is refuelled a thin wire connects the aircraft to the tanker to discharge any build-up of charge. The refuelling takes 20 minutes and a charge of 0.12 mC passes through the wire. Calculate the average current.

Magnets and electromagnetism

- ➤ There is a magnetic field around a magnet.
- ➤ When an electric current flows in a wire, a magnetic field is created around the wire.
- ➤ A force acts on any wire that is carrying an electric current through a magnetic field.
- ➤ Electric motors depend on electromagnetic forces.

A Magnets and magnetic fields

1 There are two kinds of magnet:

- **Permanent** magnets (bar magnets), made of **magnetised metal**, such as iron.
- **Electromagnets**, where magnetism is created by an **electric current**.

2 There is a **magnetic field** around a magnet (see right).

- The field affects other magnets and also magnetic materials such as iron and nickel.
- The **field lines** show the **direction of the forces** around the magnet.

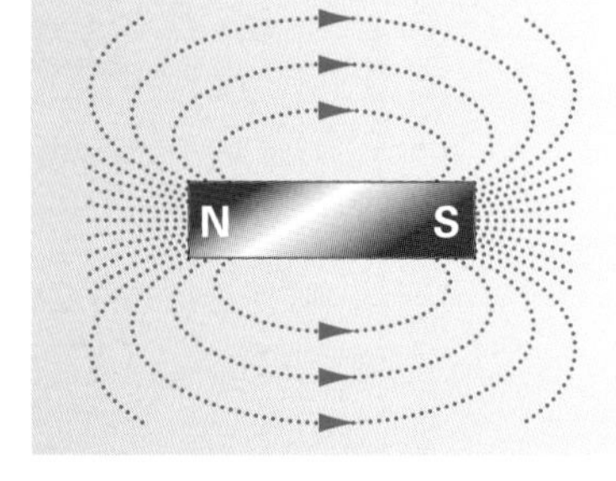

3 There is a magnetic field near a wire carrying an electric current. The field near a single wire is circular.

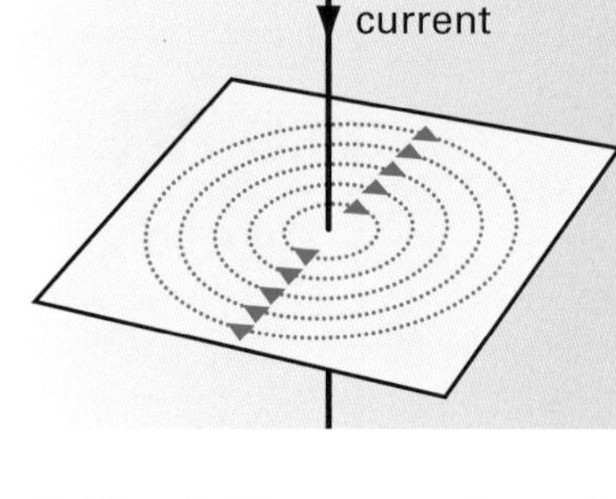

4 The field near a coil carrying an electric current (see below) looks very similar to the field near a bar magnet.

5 The field near a coil can be **made stronger** by:

- putting an iron bar through the middle of the coil
- increasing the current in the coil
- increasing the number of turns in the coil.

6 Example: How could the **direction** of the magnetic field near a coil be reversed?

Answer: The direction of the magnetic field is reversed by reversing the direction of the current in the coil.

Q How can you make the magnetic field stronger?

B Magnetic forces

When a current-carrying wire is at right-angles to a magnetic field, there is a **force** on the wire. (Reversing the current, or reversing the magnetic field, reverses the direction of this force.)

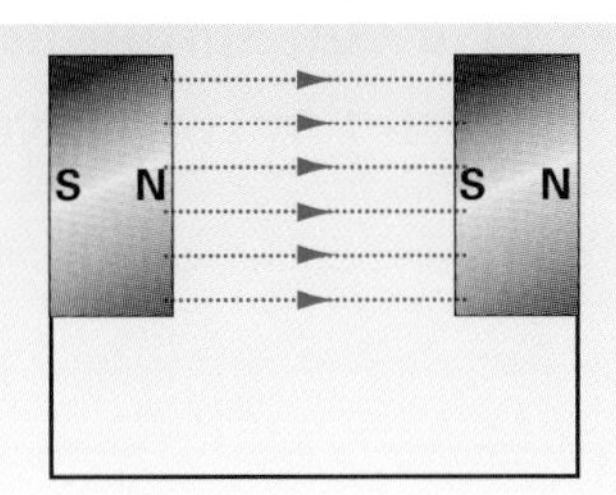

A pair of magnets produce a magnetic field in the gap between them.

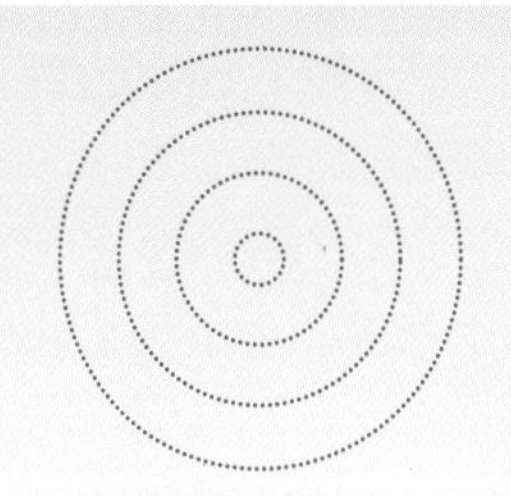

A wire carrying a current has a circular magnetic field around it.

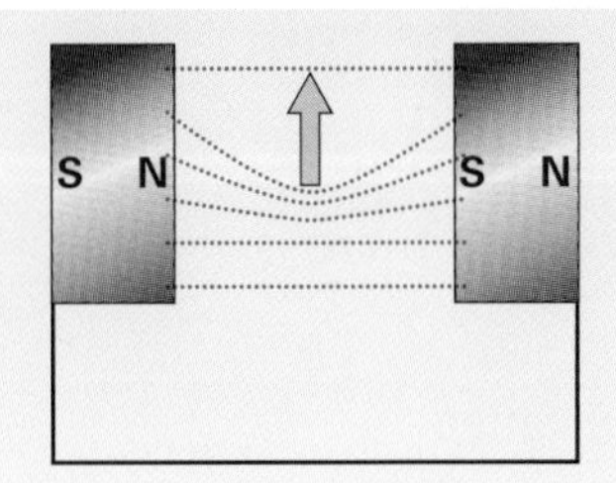

When the wire is in a magnetic field, the two magnetic fields combine to produce a 'catapult' field which pushes the wire upwards.

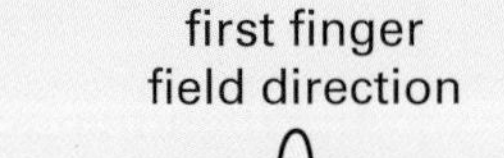

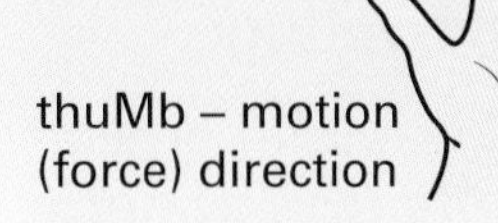

The magnetic **Field**, the **Current** in the wire and the **Force** on the wire are all at right angles.

Remember
If the magnetic field and the current are at right angles, the wire jumps in the third direction.

Q How can you make the catapult force bigger?

C Electric motor

KEY FACT

A simple electric motor has a coil of wire in a magnetic field. When a current flows the motor spins.

1. The magnetic field from the coil of wire combines with the field from the magnets.
2. The catapult field pushes the right side of the motor down and the left side up. The coil spins.
3. The **split ring commutator** makes sure that the current always passes into the side of the coil on the right – making the right-hand side always go down, and keeping the coil spinning.

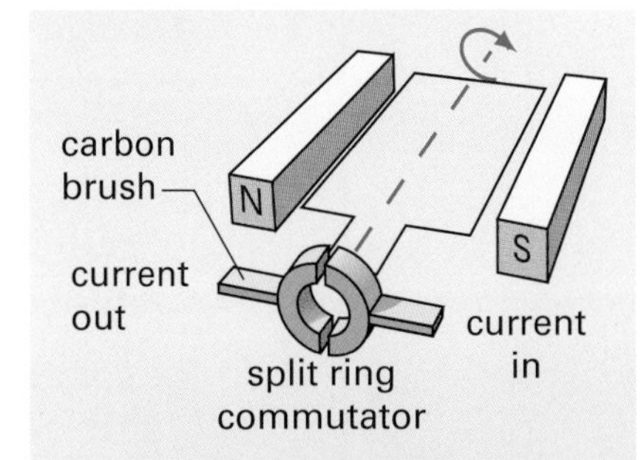

Q Explain why one side of the coil moves down and the other side moves up.

PRACTICE

1. What three separate changes could you make to the simple electric motor above to make it spin faster?
2. What two separate changes could be made to the motor to make it spin in the opposite direction?

Each separate change in direction – of current or field – will change the direction of the force.

Electromagnetic induction

- Moving a magnet into a coil produces a voltage across the ends of the coil. This is called electromagnetic induction.
- An electric generator consists of a coil spinning inside a magnetic field or a magnet spinning between coils.

A Principles of induction

KEY FACT

1 Moving a magnet into a <u>coil</u> produces a <u>voltage</u> across the ends of the coil.

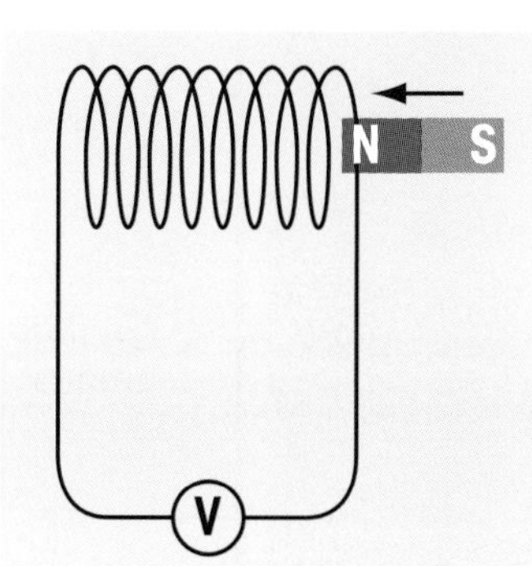

- When the magnet is pulled out, the voltage is reversed. If the magnet is stationary there is no voltage.

2 This way of generating a voltage is called electromagnetic induction.

Q Why is there no voltage induced when the magnet is still?

3 **Example:** Suggest two ways in which the voltage induced across the coil could be increased.

Answer: The voltage increases as the speed at which the coil is moved increases. The voltage is also greater if a stronger magnet is used.

B Generators

KEY FACT

1 An electric generator consists of a coil spinning inside a magnetic field or a magnet spinning between coils.

2 Replace the battery for an electric motor with a voltmeter and you have a generator.

3 As the coil turns, the wires on one side are moving up through the magnetic field at the same time as the wires on the other side are moving down.

4 A voltage is induced across the ends of the coil which are touched by the brushes.

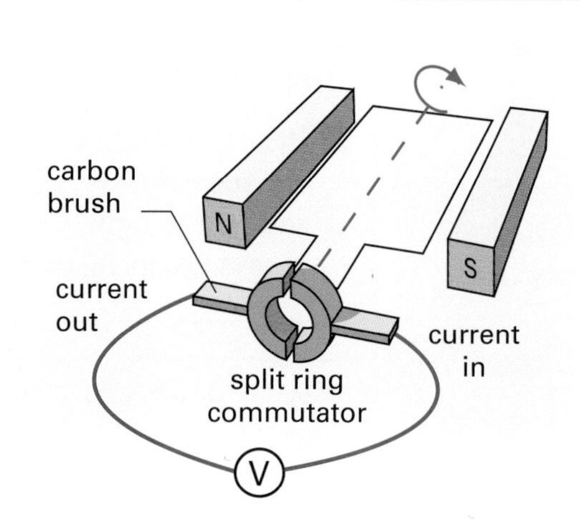

B

KEY FACT

5 Electricity can be generated by rotating a magnet inside a coil or by rotating a coil in a magnetic field.

6 An **a.c. generator** has **slip rings** so the output from one side of the coil always goes to the same brush.

The output from the a.c. generator changes every half turn of the coil and **alternating current** is produced.

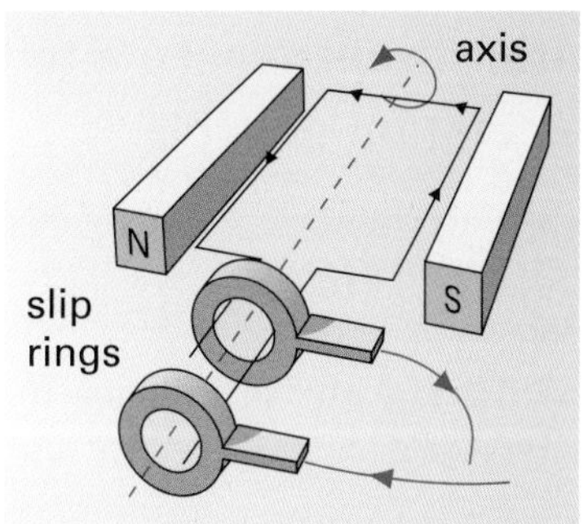

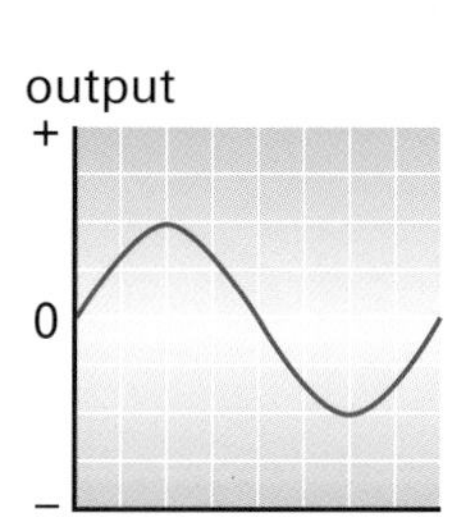

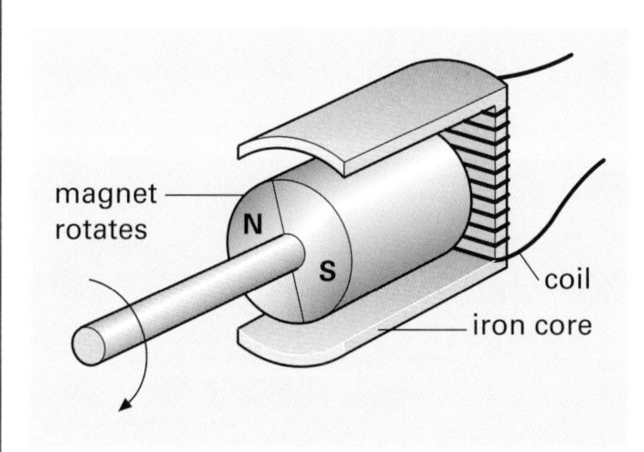

7 A bicycle dynamo is driven by the wheel of the bike.

- The dynamo spins a magnet inside an iron core. A changing magnetic field in the iron core induces a voltage across the coil.

8 **Example:** Suggest three ways in which the voltage output from the generator could be increased.

Answer: The voltage can be increased by:
- spinning the coil faster
- using a stronger magnet
- having more turns of wire on the coil.

Q What effect does spinning the coil faster have on the output voltage?

PRACTICE

1 The north end of a magnet is pushed into the right-hand side of a coil and the meter needle flicks to the right.

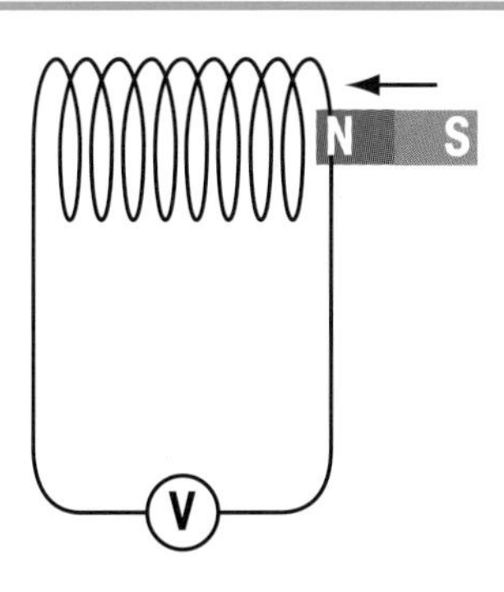

Explain what happens when each of the following occurs:

a) the north end of the magnet is pulled out of the coil

b) the north end of the magnet is pushed in the left-hand end of the coil

c) the south end of the magnet is pushed into the right-hand side of the coil

d) the south end of the magnet remains stationary inside the coil

e) the south end of the magnet is pulled quickly out of the right-hand side of the coil.

2 Explain why the spinning magnet in a cycle dynamo is like the magnet being pushed in and out of the coil.

Transformers and power stations

- Electricity is generated in power stations by spinning electromagnets near coils of wire.
- Transformers need alternating current to make their operation possible.

A Transformers

1 A transformer is a device for changing a high-voltage supply into a low-voltage one, or vice versa.

KEY FACT

2 A transformer needs an alternating current to create a changing magnetic field to induce a voltage in the secondary coil.

3 The primary coil is connected to an alternating current supply. As the current in the primary coil varies it sets up a changing magnetic field in the iron core.

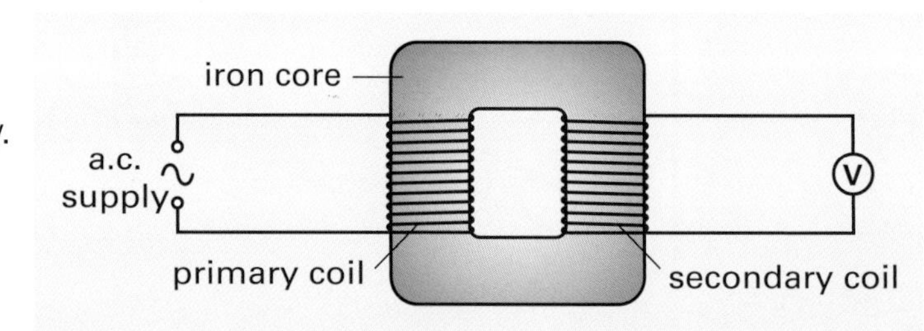

4 The changing field in the iron core induces a changing voltage in the secondary coil.

5 The voltage in the secondary coil depends on the number of turns on the coil. The bigger the number of turns on the secondary coil, the bigger the voltage across the coil.

V_p = voltage across primary coil, V_s = voltage across secondary coil

N_p = number of turns on primary coil, N_s = number of turns on secondary coil

$$\frac{V_p}{V_s} = \frac{N_p}{N_s}$$

The input power to the transformer $P_{in} = V_p I_p$.
The output power from the transformer $P_{out} = V_s I_s$.
If the transformer is 100% efficient, the power input to the transformer will equal the power output.

6 **Example:** A transformer has 200 turns on the primary coil and 400 turns on the secondary coil. The input voltage is 30 V. Calculate the output voltage.

Answer:

$$\frac{V_p}{V_s} = \frac{N_p}{N_s} \qquad \frac{30\text{ V}}{V_s} = \frac{200}{400} = 0.5 \qquad V_s = \frac{30\text{ V}}{0.5} = 60\text{ V}$$

Remember
There has to be a changing magnetic field near a wire or a wire moving in a magnetic field for electromagnetic induction to happen.

Q Why does a transformer need an alternating current?

B Generation and transmission of electricity

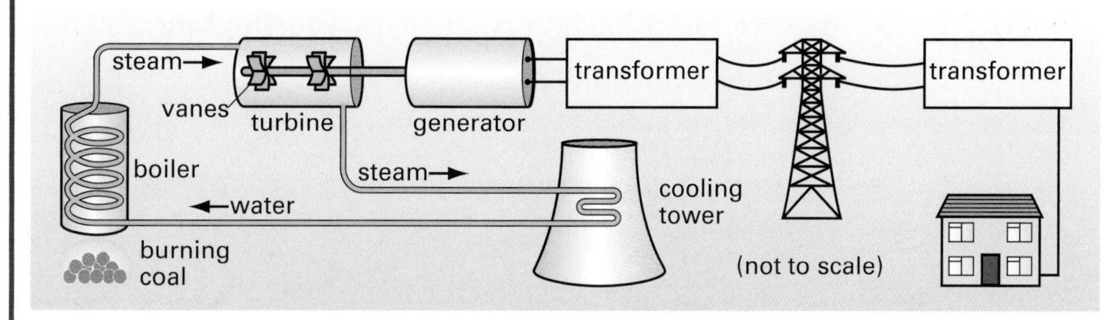

1 Much of Britain's electricity is generated in coal-burning power stations. The energy from the coal is used to boil water to produce steam. The steam is used to drive turbines, which turn generators to produce electricity.

2 Electricity is transmitted around the country on the National Grid - a network of high-voltage cables. Although the cables are good conductors, they are very long and have some resistance. As current passes through the cables, they are warmed and energy is lost.

3 Transformers at the power station step up the ac electricity to a high voltage for transmission on the National Grid.

- The higher the voltage, the lower the current to transmit the same power.
- Because the current is low there is less energy lost in heating the power lines. This is more efficient - more energy reaches the customers.

4 Local step-down transformers reduce the voltage to a safer level for use by customers.

5 **Example:** Why is the current lower, if the voltage is higher, to transmit the same power?

Answer: The transformers at the power stations step up the voltage, power = $V \times I$. If the transformers are 100% efficient, the power output will equal the power input, so $V_pI_p = V_sI_s$. If V_s increases, then I_s decreases.

Q Why does mains electricity have to be alternating current (a.c.)?

PRACTICE

1 A model train set uses a transformer to change the mains voltage to 12 V for the electric trains. What kind of transformer is this? Suggest why the voltage is reduced for the train set.

2 A 12 V 0.5 A lamp is to be used with a 230 V mains supply. The transformer has a primary coil with 460 turns.
a) How many turns will there be on the secondary coil?
b) Assuming the transformer is 100% efficient, calculate the current in the secondary coil when the lamp is lit.

3 A power station is only about 35% efficient – most of the energy from the burning fuel is lost before the electricity reaches the power lines. Suggest where energy is lost in the power station.

Distance, speed, acceleration

THE BARE BONES

- ➤ The speed of a moving object tells you the rate at which it moves.
- ➤ The velocity of a moving object tells you the speed and the direction.
- ➤ Acceleration tells you how much the velocity changes each second.
- ➤ Drawing graphs can provide information about speeds, velocities, and accelerations.

A Distance and speed

KEY FACT

1 If an object moves in a straight line, the average speed of the object can be worked out from the distance travelled and the time taken.

$$\text{speed (m/s)} = \frac{\text{distance travelled (m)}}{\text{time taken (s)}}$$

Remember
When calculating speeds use time in seconds.

Remember
You need to be able to calculate speeds from distance-time graphs.

2 A **distance-time graph** shows how far something goes in a certain time:

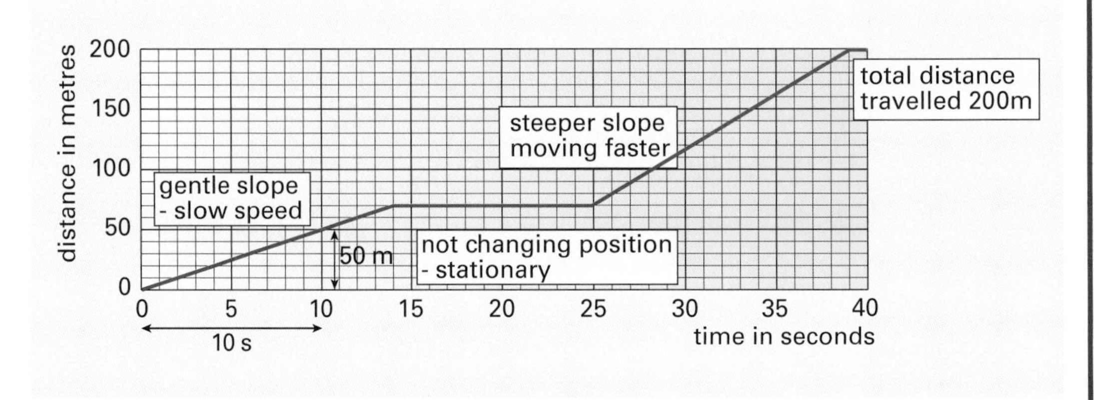

- The **gradient** of the graph gives the speed of the object.
- Between 0 and 10s the object moved from 0 to 50 m:

$$\text{speed} = \text{gradient} = \frac{50\text{m} - 0\text{m}}{10\text{s} - 0\text{s}} = 5 \text{ m/s}$$

3 Example: A boy on a bicycle travels 300 m in one minute.
What is his average speed?

Answer:

$$\text{speed m/s} = \frac{\text{distance travelled (m)}}{\text{time taken (s)}} = \frac{300 \text{ m}}{60 \text{ s}} = 5 \text{ m/s}$$

Q How far would a car moving at 30 m/s travel in 20 minutes?

B Velocity

KEY FACT

The <u>velocity</u> of a moving object tells you the speed <u>and the direction</u>.

When the space shuttle is in orbit around the Earth, it is travelling at a constant speed, but it is changing direction all the time. Its velocity is constantly changing.

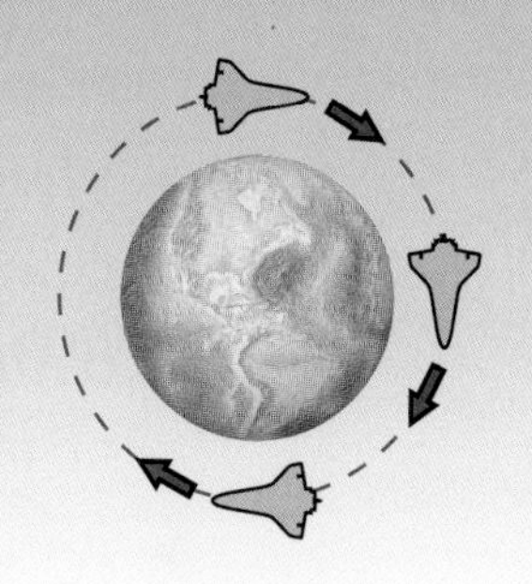

Q Explain how velocity and speed differ?

C Acceleration

KEY FACT

1 Acceleration describes how much the velocity changes each second.

$$\text{acceleration (m/s}^2) = \frac{\text{change in velocity (m/s)}}{\text{time taken (s)}}$$

Remember
You need to be able to recall and use the equations for speed and acceleration.

Remember
You need to be able to calculate accelerations and distances from speed-time graphs.

2 A **velocity–time graph** tells a story of **how the velocity of a bus changes** during a journey:

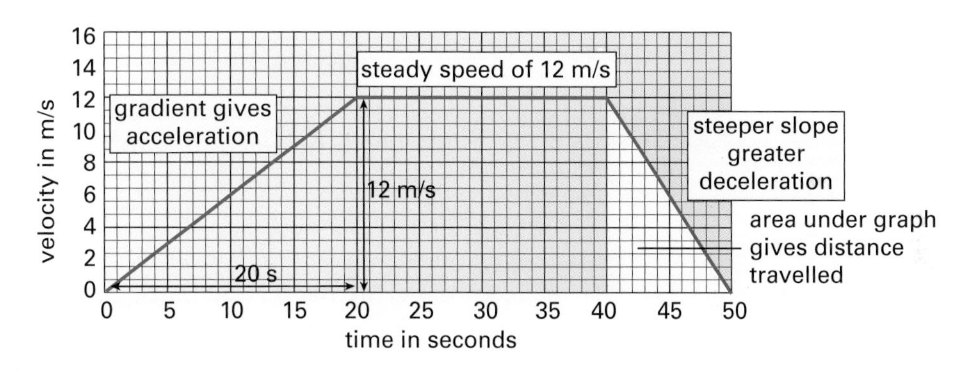

- The gradient of the graph gives the acceleration of the bus.

$$\text{acceleration} = \text{gradient} = \frac{12\text{ m/s} - 0\text{ m/s}}{20\text{s} - 0\text{s}} = 0.6\text{ m/s}^2$$

- The area under the graph gives the distance travelled by the bus.

distance travelled while braking = area of shaded triangle

$$\text{area} = \tfrac{1}{2} \times \text{base} \times \text{height}$$
$$= \tfrac{1}{2} \times 10\text{ s} \times 12\text{ m/s}$$
$$= 60\text{ m}$$

Q During lift-off, the shuttle accelerates from rest to 8400 m/s in 8 minutes. Calculate the acceleration.

Look carefully at the units on the axes when interpreting graphs.

1 A boy runs 400 m in 80 s.
a) What is his average speed?
b) How do you know that he ran faster than this during the race?

2 A mini car starts a race by accelerating from 0 to 30 m/s in 10 s. What is the acceleration of the car?

Forces

THE BARE BONES

- ➤ When two bodies interact, the forces they exert on each other are equal and opposite.
- ➤ If the forces on an object are balanced, the object remains at rest or moves at a constant speed in a straight line.
- ➤ An unbalanced force on an object causes it to accelerate.
- ➤ Force is measured in newtons (N).

A Forces come in pairs

KEY FACT

1 Whenever two objects interact they exert equal and opposite forces on each other.

- When you sit on a chair you compress the springs in the cushion. The force from the springs pushes up on you as your weight pushes down to compress the springs.

  If the chair did not push back you would fall through the chair on to the floor!

2 **Example:** What is the force that makes a sprinter accelerate away from the starting blocks?

Answer: The sprinter pushes back on the block. The bonds between the atoms in the block are compressed – the atoms get closer together, just like the springs in the cushion. The particles in the block push forward on the sprinter – the harder he pushes back, the harder the block pushes forwards.

Q What can you say about the forces acting on an object moving at a constant speed?

B Balanced forces

If the forces on an object are **balanced**, there will be **no effect on the movement** of an object.

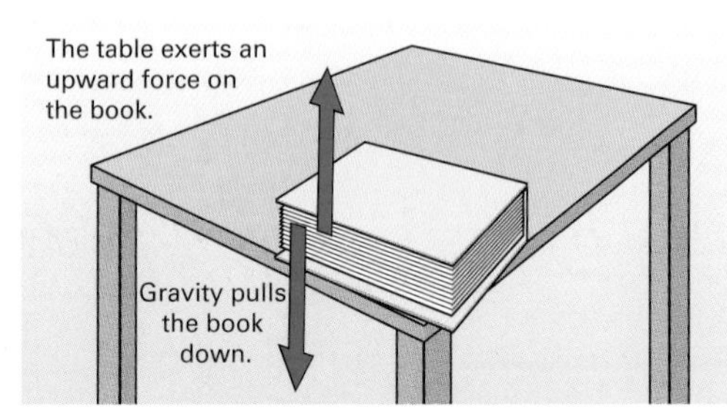

The forces are balanced - the book is stationary

The car engine drives the car forwards; resistive force act in the opposite direction.

the forces are balanced - the car continues to move at a steady speed

- Similarly, if a bicycle is moving at a steady speed along a road, the driving forces of the cyclist's legs are balanced by the drag forces acting against the motion. (*Read more about drag forces on page 146.*)

Q What are the forces that act on a book resting on a table.

C Unbalanced forces

Remember
In forces diagrams, the size of the force arrow is related to the size of the force.

If the forces on an object are unbalanced, the velocity of the object will change:

Unbalanced forces make the engine start to move in the direction of the force.

Unbalanced forces make the aircraft slow down (decelerate).

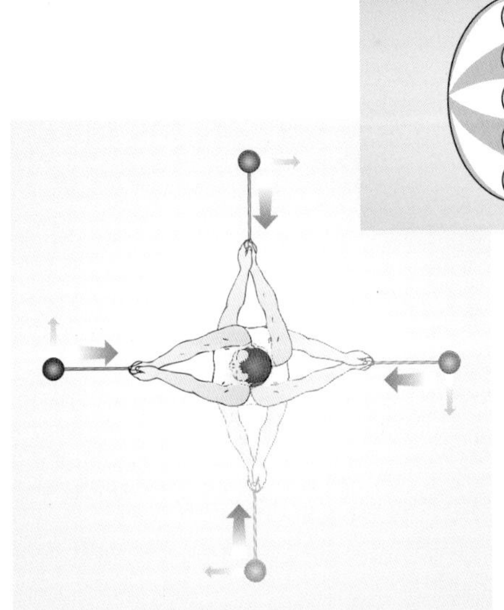

Force at right-angles to the direction of travel makes the hammer change direction.

Q What can you say about the forces acting on an object that is accelerating?

When answering questions about forces, draw a diagram showing force arrows to help you understand the situation.

D Force, mass and acceleration

1 When an unbalanced force makes an object accelerate:

- the bigger the **force**, the **bigger** the acceleration
- the bigger the **mass**, the **smaller** the acceleration.

KEY FACT

2 force (N) = mass (kg) × acceleration (m/s^2)

$$F = ma$$

3 Force is measured in **newtons (N)**. One newton is the force needed to accelerate a mass of 1 kg by 1 m/s^2.

4 **Example:** The braking force on a car gives a deceleration of 5 m/s^2. The mass of the car is 1200 kg. Calculate the braking force.

Answer: $F = ma$ = 1200 kg × 5 m/s^2 = 6000 N

Q Calculate the force that gives a 3000 kg lorry an acceleration of 3 m/s^2.

PRACTICE

1 A car with a mass of 1200 kg accelerates from 0–15 m/s in 10 seconds.
a) Work out the acceleration of the car.
b) Work out the force needed to give this acceleration.

2 Two skaters, Jane and Chris, are on the ice together. Jane pushes Chris away. Explain why Jane moves backwards as well.

Stopping forces

- ➤ The distance a car travels while the brakes are applied depends on the braking force, mass of the car and its occupants, and its speed.
- ➤ The distance a car travels before it stops depends on the speed of the car, the driver's reactions, the braking distance and the condition of the car and the road.

A Resistive forces

KEY FACT

1 Resistive forces are the forces that resist the movement of an object over a surface or through a liquid or a gas. They are also called frictional forces or drag forces.

2 Resistive forces often result in objects **warming up** as they are slowed down.

- The space shuttle has insulation to prevent the inside of the orbiter becoming overheated as the spacecraft returns through the atmosphere to Earth. *(See page 148 for more about air resistance.)*

3 **Example:** A Grand Prix car uses smooth tyres called 'slicks' on a dry track. The tyre softens as it gets warm with running. Explain why the tyre might be good on a dry track but no good on a wet track.

Answer: The soft tyre sticks well to the dry track, creating good grip. On a wet track the smooth tyre would skid across the surface. A grooved tyre provides better grip - the water can escape through the grooves.

Q Can you think of some resistive forces that will slow a car down?

B Thinking distance

KEY FACT

1 Thinking distance is the distance the car travels between the driver realising he has to put the brakes on and the brakes actually going on.

2 The thinking distance depends on the reaction time of the driver. The reaction time of the driver is affected by:

- tiredness - sleepiness is a common cause of motorway accidents
- drugs or alcohol - drugs and alcohol slow down a driver's reactions
- poor visibility - a foggy day makes it harder to see what is ahead
- distractions in the car - such as talking on a mobile phone or noisy children.

3 **Example:** Why is it dangerous to drive after drinking alcohol or taking drugs?

Answer: Drugs or alcohol make drivers less alert - they will take longer to notice a hazard and to apply the brakes. While they are reacting, the car will still be travelling towards the hazard.

Q What are the factors that would slow down a driver's reactions?

C Braking distance

The **braking distance** of a car is the distance it travels between the **brakes being applied** and the car **coming to a halt**. There are several factors that will affect how far it will travel in that time:

Remember
You should be able to use the relationship F=ma in your explanations about stopping distances.

1 The **mass of the vehicle** is important - the more massive the car, the smaller the deceleration, so the greater the distance it travels before stopping. (Similarly, a lorry loaded with bricks will take longer to stop than an empty lorry.)

2 The **speed** of the car is important - the faster the car is going, the longer time it is going to take to decelerate, so the greater the distance it travels before stopping.

3 The **force from the brakes** is important - the greater the force from the brakes, the greater the deceleration, shorter the braking time and the shorter the distance it travels before stopping. Poorly maintained brakes make it harder to stop.

Q What are the factors that affect the braking distance of a car?

4 The **road conditions** are important - on a wet road there will be less friction between the tyres and the road - the resistive forces are less and the braking distance is longer. If the driver brakes too hard, the car may skid.

D Stopping

You should be able to describe the range of factors that affect car stopping distances.

The **overall stopping distance** of a car depends on the time it takes for the driver to react and the distance the car travels while braking:

overall stopping distance = thinking distance + braking distance

Q What are the factors that affect the overall stopping distance of a car?

1 Write down what effect each of the following situations has on the stopping distance of a car, and explain why.
a) The car is heavily loaded. b) The car is moving quickly. c) The road is wet.

2 A car is travelling at 30 m/s. The driver has a reaction time of 0.5 s.
a) Calculate the thinking distance - how far the car travels between the driver seeing he needs to brake and applying the brakes.

The braking distance for this car to brake from 30 to 0 m/s is 64 m.
b) What is the total stopping distance?

Falling

THE BARE BONES

- Objects fall because gravity pulls them.
- When objects fall through air, there is also air resistance acting on them.

A Gravity

1 A gravitational force is felt near any large, massive object, such as the Sun, Moon, Earth or other planets (see page 164).

2 **Gravity** is the force that **pulls falling objects to the ground**.

3 The force of gravity on an object, its **weight**, depends on two things:

- the mass of the object, m
- the strength of the gravitational field, g

KEY FACT

4 Weight is measured in newtons:
weight (N) = mass (kg) × gravitational field strength (N/kg)

$$W = mg$$

5 At the **Earth's** surface, the **gravitational field strength** (g) is approximately **10 N/kg**. This means that the weight of 1 kilogram of matter is 10 newtons.

6 The force of gravity gives the object an **acceleration**. We call this the **acceleration due to gravity** g. On Earth $g = 9.8$ m/s^2. (You can usually use $g = 10$ m/s^2 in calculations.)

7 **Example:** Anna weighs 60 kg. What is her weight in newtons?

Answer: $W = mg = 60 \text{ kg} \times 10 \text{ N/kg}$
$= 600 \text{ N}$

Remember to use the units of newtons for weight and kilograms for mass.

Q What is your weight in newtons on the Earth?

B Air resistance

1 When objects fall through air, **air resistance** acts on them.

KEY FACT

2 The amount of air resistance depends on how fast the object is falling and on its shape.

- The bigger the **surface area** hitting the air, the bigger the air resistance.
- The **faster** the object moves, the greater the air resistance:

3 **Example:** Explain why car designers try to design cars with low air resistance.

Answer: As cars get faster, the resistive forces increase. A car with lower air resistance needs a smaller driving force, so it uses less fuel.

Q What are the factors that affect the air resistance of a moving object?

C Terminal velocity

1 A falling object in air eventually reaches a **terminal velocity**:

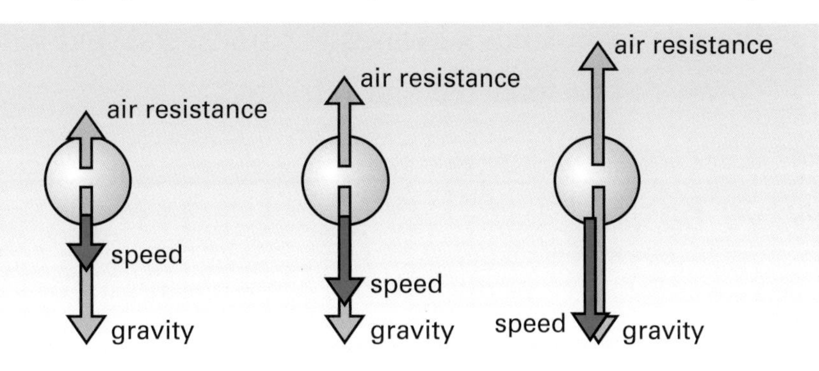

- As a falling object speeds up its air resistance increases.
- Eventually, if it falls fast enough, the pull of gravity (its weight) is balanced by the air resistance.
- It stops accelerating and falls at a steady speed - the **terminal velocity**.

KEY FACT

2 **The bigger the weight of the object, the faster it has to be falling to reach terminal velocity.**

Example: Why would a mouse land safely when falling from a first floor window, but you would break a leg - at least?

Answer: The mouse has a much smaller weight, so it would not need to fall very fast for its air resistance to be balanced by its weight. Its terminal velocity would be much slower than yours - so it would land more gently.

Use your ideas about balanced forces, weight and mass to help your explanations about terminal velocity.

Q What are the factors that affect the terminal velocity of a falling object?

PRACTICE

1 Explain why aircraft landing on an aircraft carrier use a 'drag chute' but they do not use one when landing on an airfield.

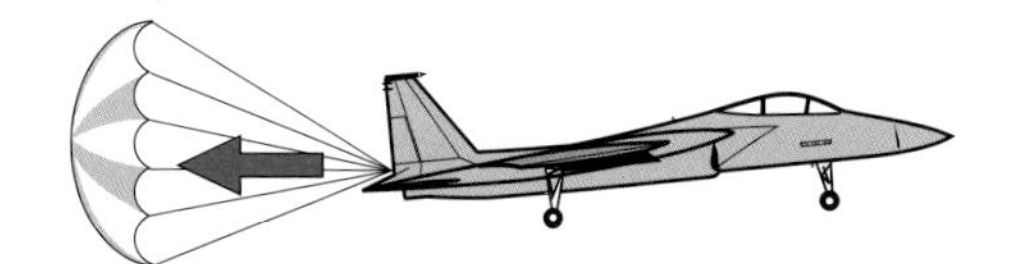

2 Astronauts visiting the Moon showed that objects dropped there fell towards the Moon. The Moon is much smaller than the Earth and so has a much weaker pull. The pull of gravity is not enough to hold on to an atmosphere. When the astronauts dropped a hammer and a feather on the Moon, they fell together - landing at exactly the same moment.

When we do the same experiment on Earth the result is different. Which lands first - the hammer or the feather? Why?

Properties of waves

THE BARE BONES

- Waves transfer energy without transferring matter.
- The two main kinds of waves are transverse and longitudinal.
- Waves can be reflected and refracted.

A Wave basics

1 Waves transfer energy without transferring matter.

2 There are two main kinds of waves, transverse and longitudinal.

KEY FACT

3 In <u>transverse waves</u> the oscillation is <u>at right-angles to the direction in which the energy travels</u>.

- Waves along a rope, light waves and waves on water are all examples of transverse waves.

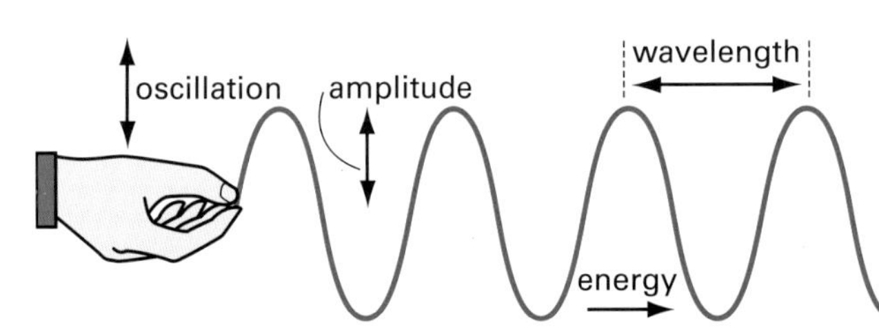

- When one end of a rope is moved up and down, the rest of the rope begins to move up and down and energy is carried along the rope by a wave.

Remember
You need to know the meaning of wavelength, frequency, wave speed and amplitude.

Remember
You need to be able to recall and use $v = f\lambda$.

- Wavelength is the distance from the crest of one wave to the crest of the next wave.
- Amplitude is the distance from the crest of a wave to the place where there is no displacement.

KEY FACT

4 In <u>longitudinal waves</u>, the oscillation is <u>in the same direction as the direction in which the energy is carried</u>.

When a slinky spring is pushed backwards and forwards, the rest of the spring moves in the same way.

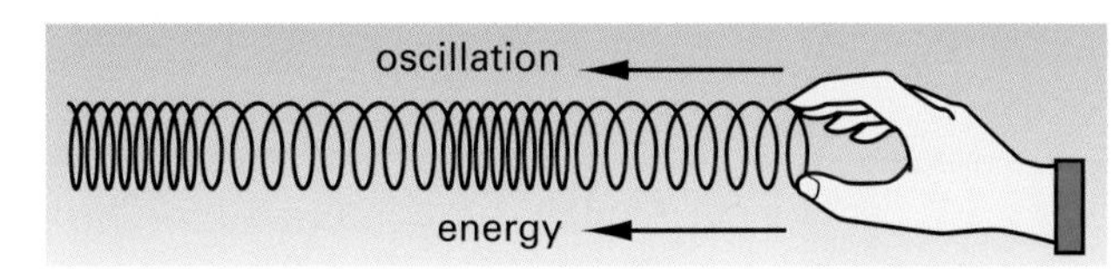

- Sound is also a longitudinal wave.

Q The 440 Hz note has a wavelength of 3.4 m in water. What is the speed of sound in water?

5 The frequency of a wave is the number of waves produced each second. The unit of frequency is hertz (Hz)

KEY FACT

6 The <u>wave equation</u> relating speed, frequency and wavelength is:

wave speed (m/s) = frequency (Hz) × wavelength (m)

$$v = f\lambda$$

(λ = the Greek letter lambda, the symbol for wavelength)

B Reflection

1 When a wave hits a barrier it 'bounces' back – it is **reflected**.

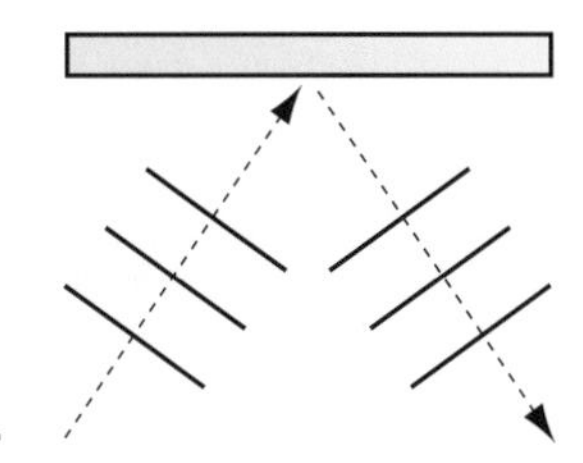

- The waves reflected from the barrier make the **same angle with the barrier** as the incoming waves.
- The wavelength and speed of the wave **remain the same**.

Q What is the rule linking the angle of incidence and the angle of reflection?

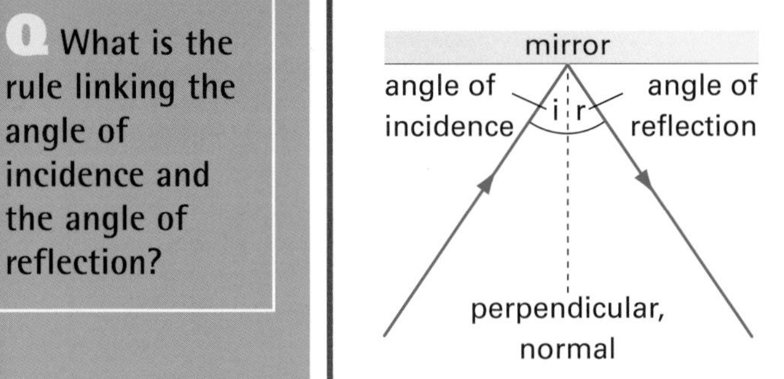

2 The same effect happens when **light is reflected from a mirror**:

- The angles of **incidence** (i) and **reflection** (r) of the light waves are measured **from a line perpendicular to the mirror**. This perpendicular line is called the **normal**.

KEY FACT

- **The angle of reflection is equal to the angle of incidence.**

C Refraction

1 When water waves pass into shallower water they are **slowed down**. This change in speed **makes the wave change direction**.

KEY FACT

- **The change in direction is called refraction.**

2 You can see the same effect when light passes **from air into glass or water**. The speed of light in glass is less than in air.

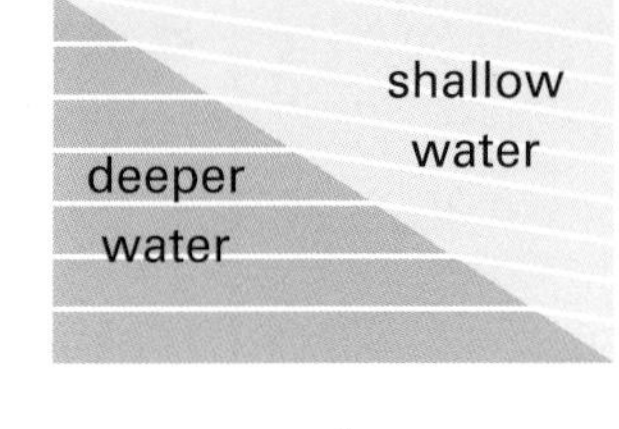

As the light slows down it changes direction. Light is refracted **towards the normal** as it is slowed down.

Q Why does refraction occur?

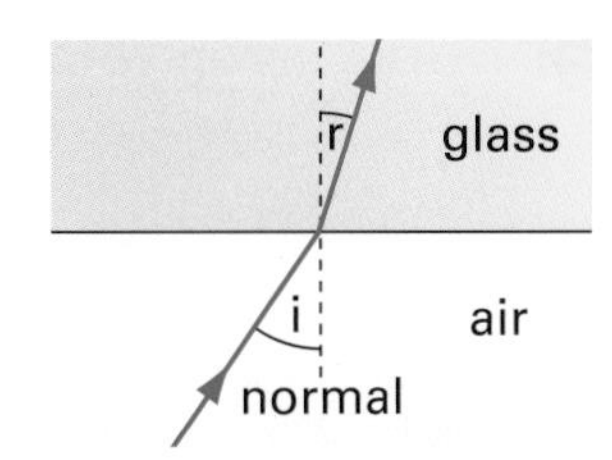

When light emerges into air it **speeds up** and changes direction again. This time it is **refracted away from the normal**.

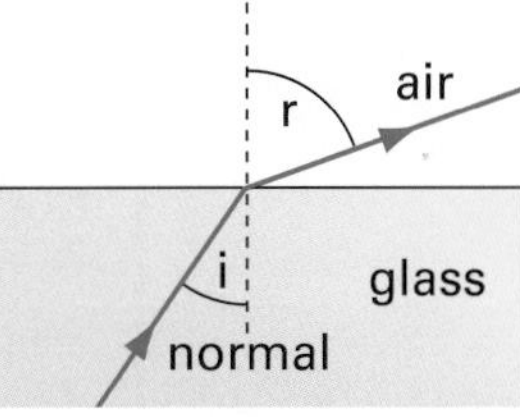

PRACTICE

1 What is the difference between transverse and longitudinal waves?

2 A musical note has a frequency of 440 Hz. The wavelength of the sound is 0.75 m. Calculate the speed of sound in air.

3 What evidence is there that sound is reflected?

THE BARE BONES

- ➤ Total internal reflection occurs when the angle of incidence inside glass or water is greater than the critical angle.
- ➤ Diffraction is the spreading out of waves when they pass through a narrow gap.

A Total internal reflection

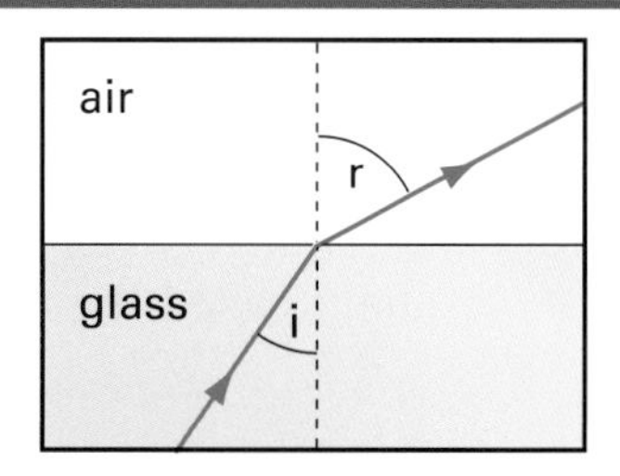

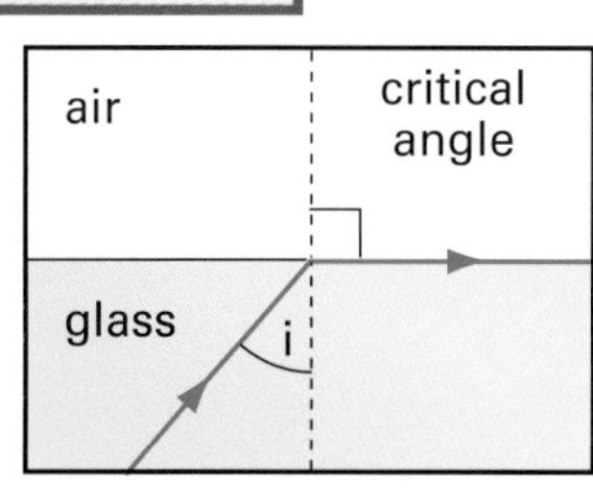

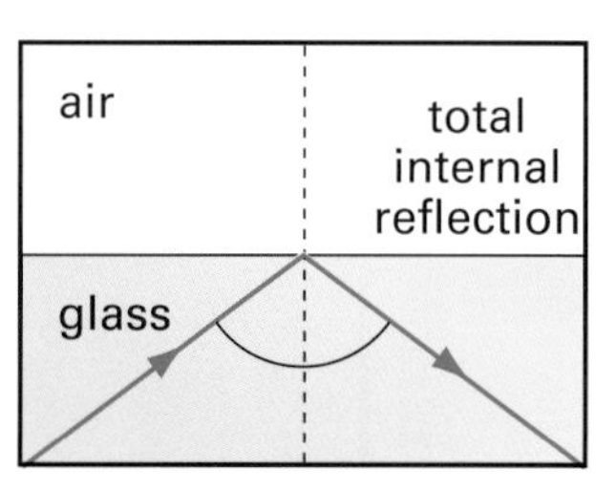

Remember
Total internal reflection occurs when the angle of incidence inside glass - or water is greater than the critical angle.

1 When light reaches the boundary between glass and air, some is reflected back into the glass, but most emerges into the air. The light emerging into the air is refracted **away** from the normal.

2 As the angle of incidence inside the glass increases, the angle of refraction outside increases until the emerging ray is passing parallel to the edge of the block. The angle of incidence when this happens is called the **critical angle**.

3 If the angle of incidence inside the block increases any further, the light cannot emerge and is all reflected back inside the block. This is **total internal reflection**.

4 **Example:** How would you complete the diagram to show how total internal reflection occurs in a right-angled prism?

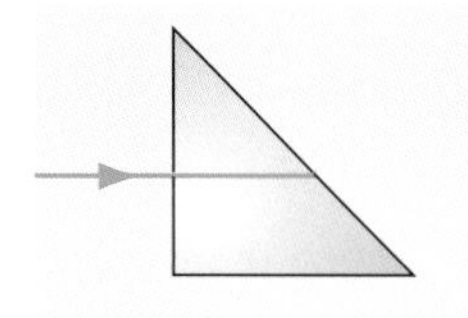

Answer:

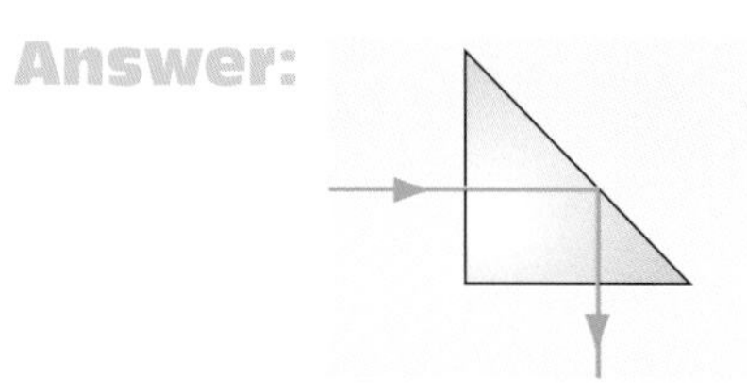

5 **Visible light** can be passed down **optical fibres**, using **total internal reflection**.

- An **optical fibre** is a thin strand of glass coated in a protective material. The light is reflected off the inner surfaces of the fibre and stays within the fibre even when the fibre bends.

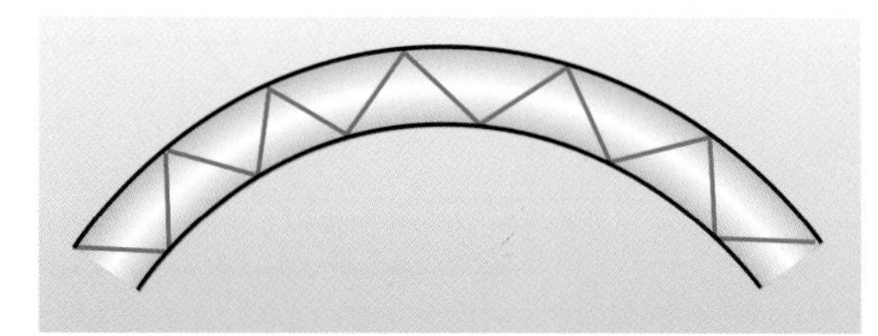

- An **endoscope** is a device that incorporates optical fibres. Doctors use it to see inside patients without the need for an operation.
- Optical fibres are also used for **communication** (see page 156).

Q How does refraction stop a fish seeing a predatory bird?

On ray diagrams use small arrows to show the direction of the ray of light.

B Diffraction

1 When waves pass through a **gap in a barrier**, they **spread out** as they pass through to the other side. This is called **diffraction**.

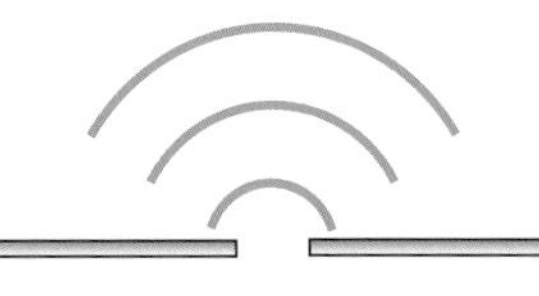

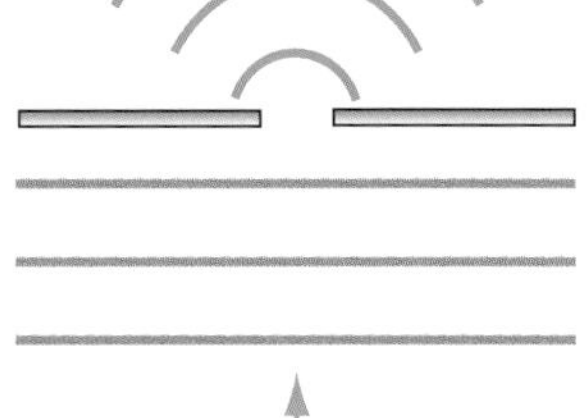

KEY FACT

2 **Diffraction happens best if the wavelength of the waves is about the same as the width of the gap.**

Waves are more curved with a narrower gap.

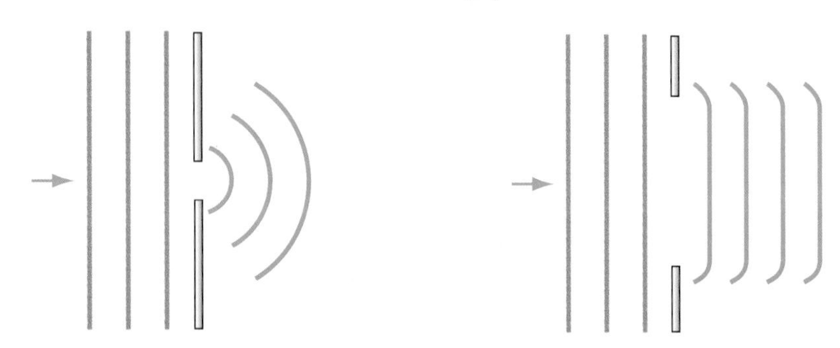

3 Diffraction provides evidence that **light**, **water waves** and **sound** all behave in the same way.

- You can hear what is being said in another room, even if the people talking are out of sight. **Sound is diffracted** as it passes through the doorway. This happens because the wavelength of the sound is about the same as the width of the doorway.
- We do not often notice the **diffraction of light** because the **wavelength of light is very small.** This means that light only diffracts noticeably when the gap is very small too.
- Diffraction of **radio waves** is useful in communications (*see page 156.*)

Q When does diffraction happen best?

1 Complete the ray diagram to show how total internal reflection is used to make these two prisms into a periscope.

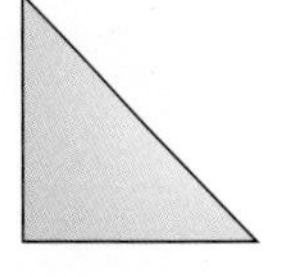

2 Explain how we know that sound waves diffract.

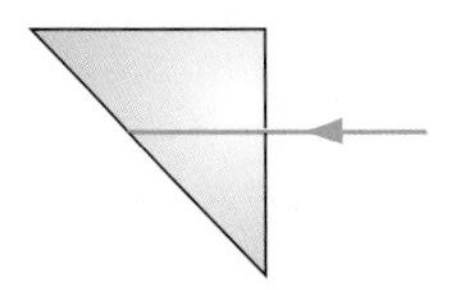

Electromagnetic waves

THE BARE BONES

- Electromagnetic waves include light waves, radio waves, infrared, ultraviolet, X-rays and gamma rays.
- These waves are all similar in nature but have different wavelengths and energies.
- Different parts of the electromagnetic spectrum have different uses and hazards.

A The electromagnetic spectrum

The **electromagnetic spectrum** is a family of waves that are able to travel through a **vacuum**, where they all travel at the **same speed** – the **speed of light**. Electromagnetic waves are **transverse waves**.

short wavelength 0.000 000 000 001 m high frequency 10^{20} Hz highest energy	gamma rays	• emitted by some radioactive materials (see page 174) • used in medicine to kill cancer cells and to trace blood flow • used to kill harmful bacteria • large doses may damage human cells
	X-rays	• pass through soft tissue but not bones or metals • used to produce shadow images of bones • large doses may damage human cells
	ultraviolet	• causes tanning of the skin • large doses may cause skin cancer
wavelength 0.000 000 5 m	violet visible light red	• detected by the eye • used for vision and photography • used through optical fibres for viewing inside the human body and other inaccessible places
	infrared	• is radiated from warm and hot objects and causes heating • used in grills, toasters, radiant heaters • used in remote-control devices, security cameras and communication
	microwaves	• some wavelengths absorbed by water; used in cooking; can damage living cells • longer wavelengths: radar/mobile phone
long wavelengths 1 m –100 000 m low frequency (104 Hz) lowest energy	radio waves	• broad range of wavelengths used in communication. • radar used to track aircraft and shipping • radar guns measure speed of cars

Q Give an example of a use for each wave in the electromagnetic spectrum.

B Some uses and hazards of electromagnetic waves

KEY FACT

1 Short wavelength electromagnetic waves such, as X-rays, gamma rays and ultraviolet, have very high energies.

- These waves can **damage living cells**.
- X-rays and gamma rays are **used to kill cancer cells**, but the radiation has to be applied **carefully** so that **healthy cells are not damaged** too. (There is more about gamma rays on pages 174–179.)

  The x-rays are sent in small doses from different directions. The healthy cells do not receive too much radiation; the tumour cells receive most.

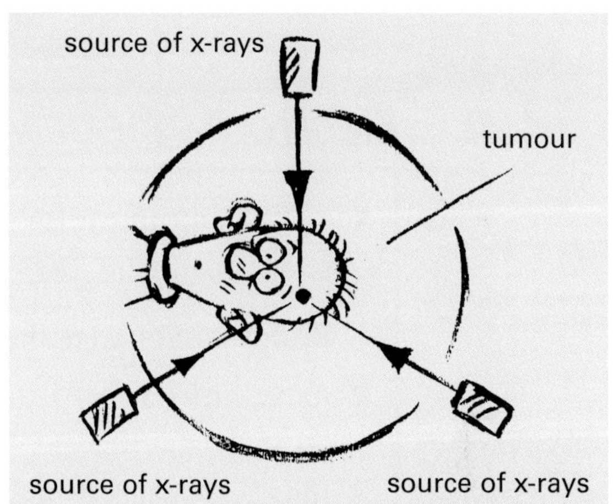

Remember
Long wavelength
- low frequency
- low energy;
Short wavelength
- high frequency
- high energy.

- Light from the Sun includes **ultraviolet waves** that can damage skin cells and lead to **skin cancer**. This is the reason you should be **careful how much you sunbathe.**

KEY FACT

2 Infrared radiation is emitted by all warm bodies, and can be detected by suitable instruments:

- **Infrared security cameras** are sensitive to infrared radiation. They can be used to 'see in the dark'.
- **Infrared detectors** are used to switch on security lights or burglar alarms if intruders are detected.

3 **Microwave ovens** produce electromagnetic waves with a wavelength of around 10 cm.

- Water, fats and sugars easily absorb these waves. The molecules vibrate more vigorously, making the food in the oven hot.

4 Microwaves of different wavelengths are also used in **communication** (*see page 156*).

- Microwaves got their name originally because they have a **shorter wavelength than other radio waves** – even though, compared with other electromagnetic radiation, their wavelength is **long**.

Q Why can exposure to too many X-rays be dangerous?

PRACTICE

1 What properties do all parts of the electromagnetic spectrum have in common?

2 Suggest one use and one hazard of each of the following electromagnetic waves:
a) X-rays
b) microwaves
c) ultraviolet waves.

You need to know some uses and hazards of the different types of radiation in the electromagnetic spectrum.

Communicating with waves

THE BARE BONES

- ➤ Electromagnetic waves used for communication include infrared, radio waves and microwaves.
- ➤ Digital communications allow more information to be carried with less interference.

A Electromagnetic waves and communication

KEY FACT

1 Electromagnetic waves can be used to transmit coded information fast over long distances.

- Radio waves were the first waves to be used in this way.

2 Today, light and infrared rays transmitted down optical fibres are important for communication, as are microwaves.

3 The electromagnetic waves used to transmit radio and television broadcasts are able to diffract around buildings, which means there is less 'shadow' where the signal does not reach.

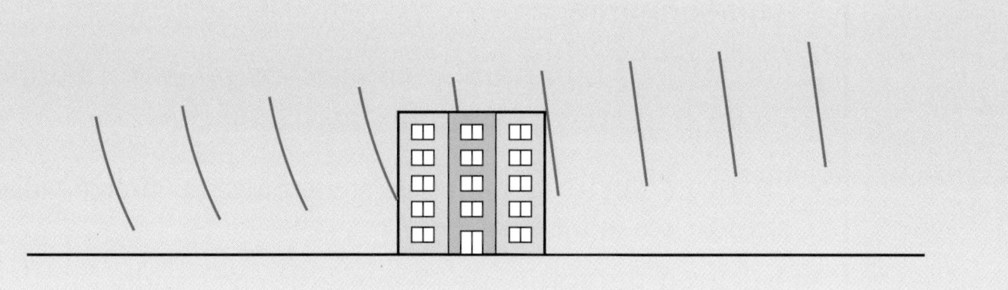

4 All signals get weaker as they travel. Booster stations receive the signal and retransmit a stronger signal.

Q Why are electromagentic waves useful for transmitting information?

B Optical fibre communication

1 Telephone companies use cables of optical fibres to transmit telephone calls, computer signals and television programmes using infrared waves. These have become an important part of the 'information revolution'.

2 An optical fibre carries far more messages than a similar thickness of copper cable. Less energy is lost along the way, and so fewer 'booster stations' are needed to overcome these energy losses.

- Infrared signals in optical fibres cannot be 'tapped' and suffer less from interference than an electrical signal in a copper cable.
- Modern optical fibres are so fine that infrared passes straight down the cable.

3 There is more about optical fibres on page 152.

Q What are the advantages of optical fibres to carry telephone calls?

C Microwave communications

1 Microwaves are used to transmit calls from mobile phones.

2 Microwaves are used to transmit telephone calls across the country from transmitter to transmitter.

3 Microwaves travel in **straight lines** and the curvature of the Earth limits the distance between transmitters to about 40 km.

4 **Satellites** are used as relay stations to send signals around the world. Microwaves carry the signals to an orbiting satellite, which receives the signal and bounces it on its way.

5 There is some uncertainty about how safe the signals are - do they warm up the brain or change some brain cells?

6 **Example:** How does the satellite television signal reach the television?

Answer: The dish on the outside of the house collects microwave signals from a satellite. The signal is converted to an electrical signal and passed down a cable to the TV.

Q Why does there need to be a microwave transmitter every 40 km?

D Digital signals

1 A conventional telephone converts the sound waves to a changing electrical signal whose pattern matches the original wave. This is called an **analogue signal**.

2 Modern telephone systems convert the analogue electrical signal to a **digital signal** – a stream of numbers that describes how the analogue signal changes.

KEY FACT

3 **Digital signals allow the optical fibre or radio wave to carry even more information, many television channels or lots of telephone conversations.**

- Digital signals are affected less by **interference**.
- E-mail and fax are both sent by digital signals, too.

4 **Computers** also use digital signals.

Q What are the advantages of using digital signals?

PRACTICE

1 What are the advantages of using infrared through optical fibres for communications?

2 Describe the difference between analogue and digital signals.

Sound and ultrasound

- Sound is carried by the particles in a medium vibrating – it cannot travel through empty space.
- The pitch of a musical note depends on its frequency of vibration.
- Ultrasound is sound at a pitch too high for humans to hear.
- Various technologies make use of ultrasound, and some animals also use it for echolocation.

A Making sounds

1 Sounds can be transmitted through solids, liquids or gases, but not through empty space.

KEY FACT

2 Sounds are made when objects vibrate.

- When a ruler vibrates, the air around the ruler is made to vibrate.
- The particles in the air alternately come closer together, creating a compression, and then spread further apart, creating a rarefaction. The sound spreads through the air.

3 The particles vibrate with the same frequency as the sound and in the same direction as the sound travels. Sound waves are longitudinal waves.

Remember
The frequency of the sound is the number of complete vibrations per second. Frequency is measured in hertz (Hz).

4 The energy of the sound causes other objects to vibrate. The sound makes your eardrums vibrate; this is how you detect sounds.

- The lower the frequency of the sound, the lower the pitch of the note.
- The bigger the amplitude of the vibration, the louder the sound. If the sound is too loud, the large vibrations may damage your ears.

5 **Example:** What effect does tightening a guitar string have on the pitch of the note?

Answer: Tightening a guitar string increases the frequency of the vibrations. The higher the frequency, the higher the pitch of the note.

If you cannot remember the relationship between frequency and pitch think of twanging a rubber band: a thick, loose band vibrates slowly and gives a low note. Tighten it and the vibrations get faster and the pitch higher.

Q What kind of wave is a sound wave?

B Ultrasound

1 Sounds that have too high a frequency for humans to hear (above about 20 000 Hz) are called **ultrasound**.

2 Ultrasound is partly reflected when it meets the boundary between two different materials. The time taken for the ultrasound signal to be reflected is measured to calculate how far away the boundary is. A computer takes many measurements to build up a picture.

- **Ultrasound** is used to look inside the body.

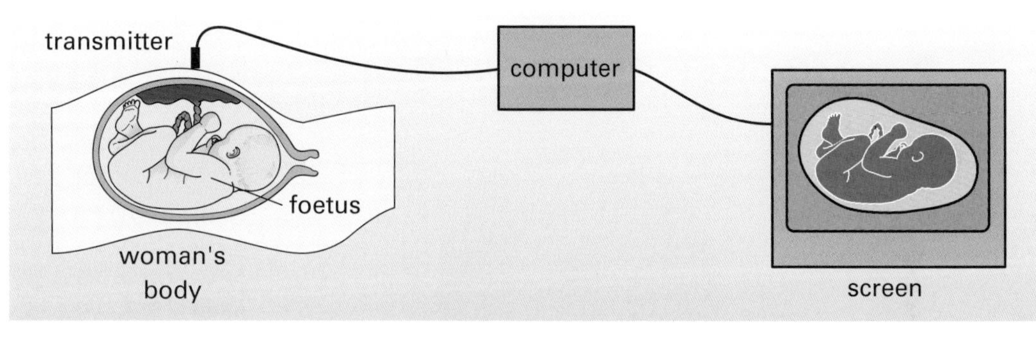

The reflected wave is used to build up a picture of the inside of the body. Ultrasound scans are used to look at babies before birth and also to look at other organs.

Remember
You need to be able to recall and use

speed = $\frac{\text{distance travelled}}{\text{time}}$

3 Bats use **echolocation** to detect objects around them. They detect the echoes from 'clicks' they emit and calculate the distance away of the object.

- Dolphins and submarines use echolocation too.

4 Ultrasound is used in industry for detecting flaws inside metal casting. A crack inside the metal will be detected as a boundary.

- Ultrasound vibrations are used for cleaning small objects. Jewellery and electronic components are placed in a cleaning fluid. Ultrasonic vibrations pass through the fluid, causing the dirt particles to shake loose.

5 **Example:** A diving vessel uses echolocation to detect submerged objects. An ultrasound beam is transmitted to the bottom of the seabed. A reflection is detected 0.6 seconds later. Sound travels at 1500 m/s in seawater. How far away is the wreck?

Remember an echo has to go there and back – it travels twice the distance in twice the time.

Answer:

distance travelled = speed × time
= 1500 m/s × 0.6 s
= 900 m – but that is <u>twice</u> the depth.
So the wreck is 450 m down.

Q Why do you think ultrasound is used to look at a baby in the womb, rather than X-rays?

PRACTICE

1 A bat sends a signal out. It is reflected from the wall of the cave 0.2 seconds later. The cave wall is 30 m away. What is the speed of sound?

2 What property of ultrasound is used to make the image of a baby?

Structure of the Earth

THE BARE BONES

- ➤ The Earth has a layered structure.
- ➤ The thin outer layer is broken into large plates that move over the surface, causing earthquakes, mountain-building, and movement of continents.
- ➤ Waves triggered by earthquakes provide evidence of the internal structure of the Earth.

A Structure of the Earth

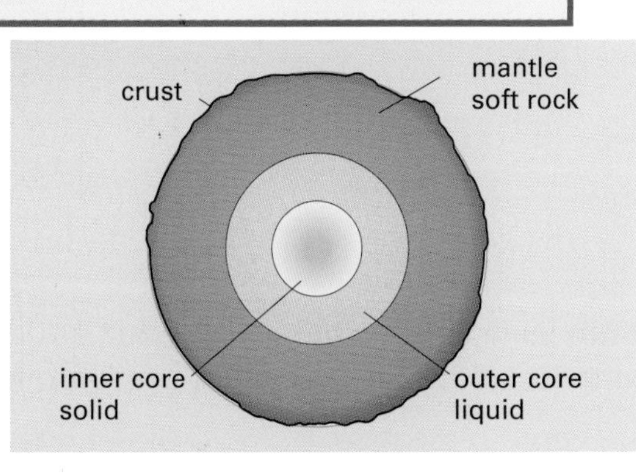

1 The Earth has a **layered** structure. The thin outer part, the **lithosphere**, is about 70 km thick.

- The lithosphere includes both the **Earth's crust** and part of the underlying **mantle**.

2 The lithosphere is broken into large sections, called **plates**, that move by slow **convection currents** deep within the Earth. The theory that describes how they do this is called **plate tectonics**.

3 The plates can move in **three different ways**. Each has different effects on the Earth:

- Where the plates **move apart**, hot molten rock rises and solidifies to form new plate material. This is happening under the Atlantic Ocean floor, where the crust is thinner.
- Where the plates **move together**, **mountains may form** by folding, **volcanoes** may erupt and **earthquakes** occur. This is happening on the west coast of South America, where the Andes mountain range has formed.
- Where plates **slide past each other**, **earthquakes** occur. This is happening along the west coast of North America in California.

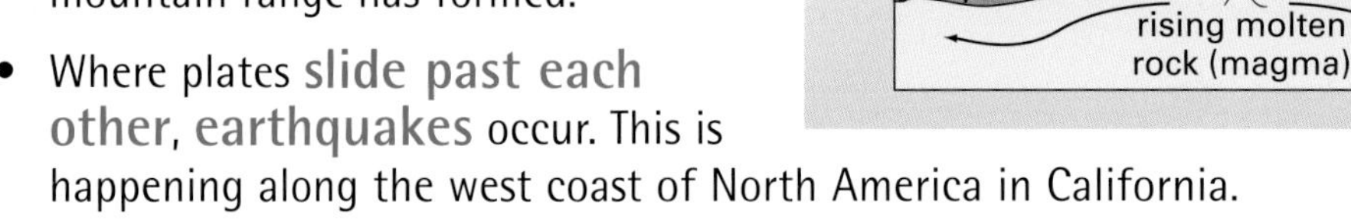

4 The **energy** required for these convection currents comes from **radioactive decay** within the Earth.

Q What are the three different ways in which plate movement can affect the Earth?

B Evidence of past plate movements

The rock record provides evidence to support plate tectonic ideas.

- Africa and South America have **closely matching coastlines.** The rocks, fossils, and some of the present-day animals they share suggest that they were once close together.

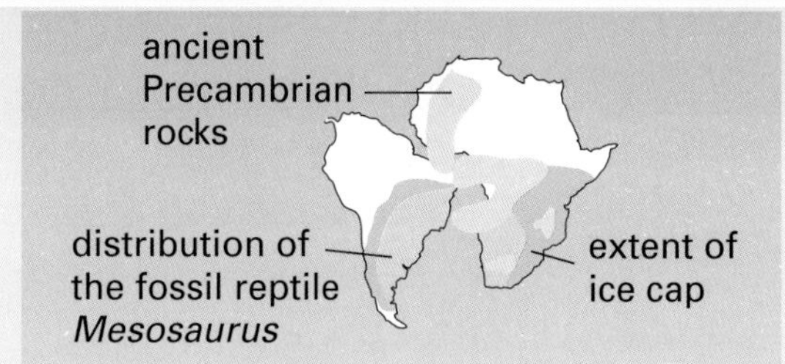

- New plate material forms under the sea along the Atlantic Ridge. As this rock solidifies it records the **direction of the Earth's magnetic field.** The direction of the Earth's magnetic field 'flips' periodically. The magnetic patterns give us information about the way the plates were formed.

Q On the map, find where the plates are moving apart to form the Atlantic ridge.

C Earthquakes and seismic waves

1. Earthquakes occur when the moving tectonic plates stick and then move suddenly. This happens at the earthquake **epicentre**.
2. The shock from large earthquakes shakes the ground surface and can destroy buildings. Some of the energy from the earthquake is carried through the Earth by **seismic waves.**
3. Seismometers are used to detect the waves.
4. Seismic **primary (P)** waves are the **fastest**: they travel at 8 to 13 km/s. They are **longitudinal waves** that can pass through solids, liquids and gases and **can pass right through the Earth.**
5. Seismic **secondary (S)** waves are slower, **transverse waves**: they travel at 4 to 7 km/s. These transverse waves can **only pass through solids.** They are stopped by the core, showing that **the outer core is liquid.**
6. The waves are **refracted** as they pass through the different types of material inside the Earth. **Seismologists** use information about the types of waves that are transmitted and how long they take to arrive at different 'listening stations' to learn more about the structure of the Earth.

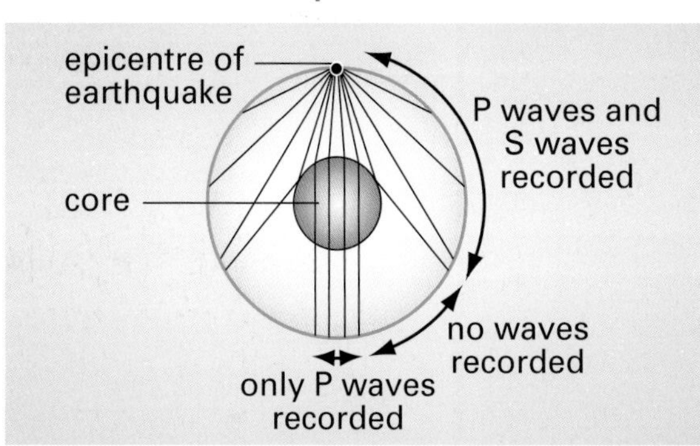

Q What are the two main types of seismic wave?

PRACTICE

1. Write down three differences between P waves and S waves.
2. An earthquake in Turkey is detected 3000 km away. Which waves, S or P will arrive first? What is the length of the time delay before waves begin to arrive?

The Earth in space

THE BARE BONES

- The solar system consists of the Sun, nine planets, the asteroid belt, comets and meteors.
- All the bodies in the solar system are held in orbit by gravity.

A The solar system

KEY FACT

1 The solar system consists of the Sun, nine planets, the asteroid belt and many comets and meteors.

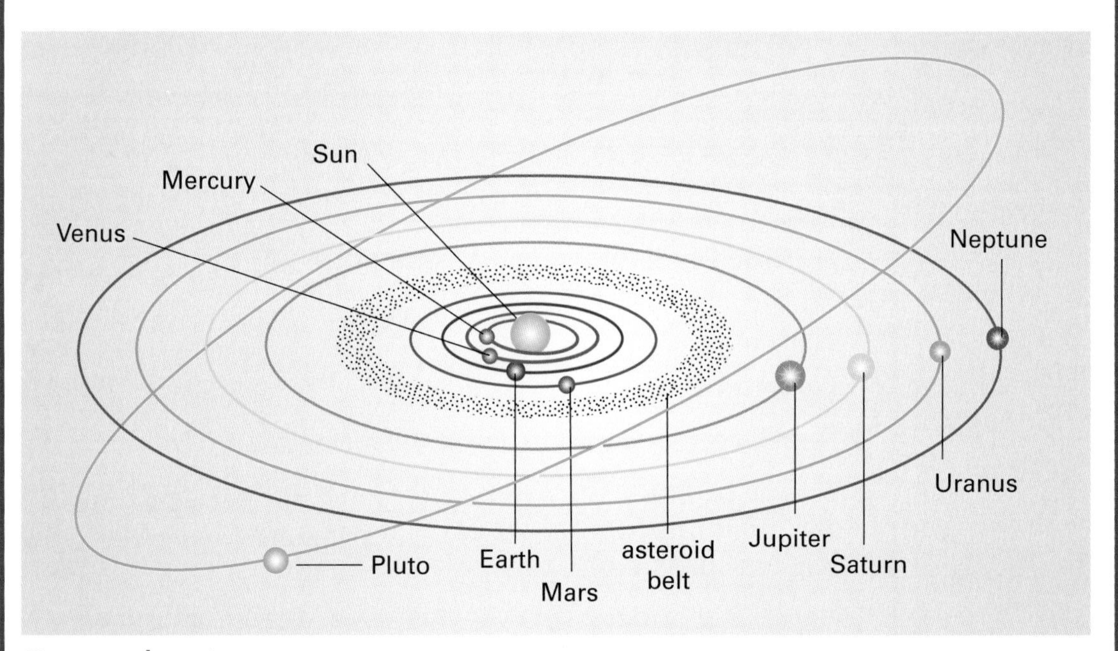

2 Nine **planets** orbit the Sun. They are different sizes, and take different times to complete their orbits:

Planet	Number natural satellites	Diameter in kilometres	Ave. distance from the Sun in millions of kilometres	Time to orbit Sun in Earth units
Mercury	0	4880	58	88 days
Venus	0	12100	108	225 days
Earth	1	12800	150	365 days
Mars	2	6800	228	687 days
Jupiter	16	143000	778	11.9 years
Saturn	18	121000	1430	29.5 years
Uranus	15	52000	2870	84.0 years
Neptune	2	49400	4500	165 years
Pluto	1	3000	5900	248 years

Q List the planets in order of distance from the Sun.

B Orbits of planets, comets and moons

1 All the planets except Pluto have **elliptical orbits in the same plane**.

- The orbit of **Pluto** is more elliptical and **at an angle** to the plane of the other planets.

2 Between the **four inner planets** and the **outer planets** lies the **asteroid belt**, made up of dust and rocks. Other asteroids orbit further from the Sun, and some come much closer to the Earth.

3 **Comets** are lumps of ice, dust and gas.

- The **orbits of comets** are very elliptical. They pass close to the Sun, when they can be seen, and then travel far beyond Pluto.
- When comets are close to the Sun they move **faster**, when they are far away they move **more slowly**.

Devise a way of recalling the order of the planets from the Sun e.g. My Very Easy Method Just Speeds Up Naming Planets. (Now you have to match the planet names to the first letter of each word!)

- When comets pass close to the Sun, **some of their ice melts**, releasing a **tail of dust and gas**, which glows in the heat of the Sun.

4 The Earth and some of the other planets have one or more **moons**. These moons are natural satellites, held in orbit by **gravity**. Earth's Moon makes one orbit of the Earth every **month**.

5 **Example:**

Why are comets are only visible when they come close to the Sun?

Answer:

When the comets come close to the Sun, they are closer to the Earth, too, so are easier to see. When the comets are a long way from the Sun they are only seen by reflected light, but when they come close to the Sun, the ice melts and the gases and dust become so hot that they glow.

Q How is the orbit of Pluto different from the other planets that orbit the Earth?

PRACTICE

1 The orbit of the asteroid Chiron is about 2050 million kilometres from the Sun. Where does that place it in the solar system? Suggest an approximate time for the asteroid to complete one orbit.

2 State a relationship between the length of time a planet takes to orbit the Sun and the distance the planet is from the Sun.

Moving in orbit

THE BARE BONES

- ➤ Gravity attracts all masses.
- ➤ Artificial satellites orbit the Earth.

A Gravity

KEY FACT

1 All masses are attracted to all other masses by the force of gravity.

2 The gravitational force between two bodies depends on the mass of each of the bodies and how close together they are. The force of gravity is only noticeable if one of the masses is very big - like the Earth or Moon.

At the surface of the Moon, the gravitational field strength is about 1.6 N/kg. This means that 1 kilogram of matter would only weigh 1.6 newtons on the Moon. The low gravity is the reason that the astronauts seemed to walk with a bounce in their step.

3 All the bodies orbiting the Sun are held in orbit by the force of gravity.

4 The time taken for a planet to orbit the Sun is its year.

5 The further a planet is from the Sun, the weaker the pull of gravity on it; the longer its orbit, the longer its year.

6 **Example:** Why do planets further from the Sun take longer to orbit the Sun?

Answer: The distance round the orbit is greater, so the orbit takes longer. The pull of gravity is weaker for the distant planets, so the planet moves more slowly.

Remember
You need to appreciate the difference between the mass of an object, measured in kilograms and the weight of an object, a force measured in newtons.

Q What are the two factors that determine how strong gravity is?

B Artificial satellites

Monitoring satellites study what is happening on the Earth:

Monitoring satellites in low orbits over the North and South Poles scan the Earth each day. They are used to observe the weather, land-use across the world, and the movement of armies. The low orbits allow the satellites to obtain better images of the Earth. Also, as they orbit the Earth, the Earth turns beneath them, so over a number of days they will 'see' the whole surface of the Earth.

Communications satellites are put into a high orbit around the Earth:

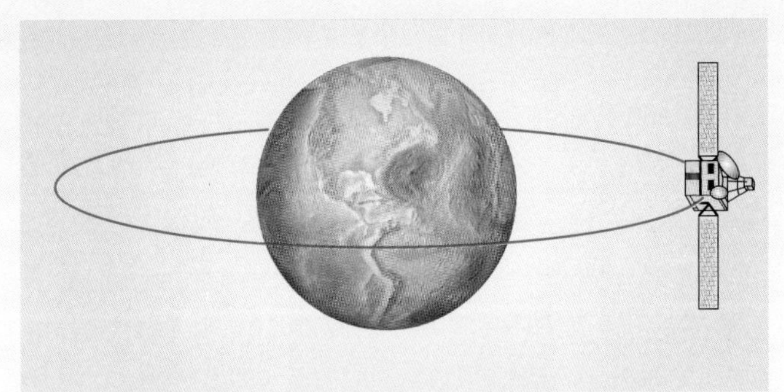

At 36 000 km above the Earth, a satellite orbits around the Earth at exactly the same rate at which the Earth itself turns. So it **appears to remain in one position** above the Earth. This is called a **geostationary** orbit, and allows the satellite to stay in the same position relative to signalling stations on the ground.

Artificial satellites are put into orbit around the Earth for various purposes. They are held in orbit by gravity. To stay in orbit **at a particular distance** the satellite must move **at a particular speed**. Satellites **close** to the Earth orbit **much more quickly** than those in higher orbits.

Other satellites, such as the **Hubble space telescope**, are designed to make observations of the **rest of the solar system** and **deep into the universe**. Hubble is 600 km above the Earth.

Q List three different uses of satellites.

PRACTICE

1. Anna has a mass of 50kg. On Earth, the gravitational field strength is about 10 N/kg. On the Moon, the gravitational field strength is about 1.6 N/kg. Calculate her weight on:
   a) the Earth, and
   b) the Moon.
2. The asteroid Eros is about 33 kilometres long, 13 kilometres wide and 13 kilometres thick. It is said that a basketball player with a 1 metre vertical leap on Earth could jump nearly 2 kilometres high and risk putting himself in orbit! Explain why this might happen.

THE BARE BONES

- The Sun and other stars have a life cycle – they evolve from one kind of star into another.
- The Milky Way galaxy is one of many millions of galaxies in the Universe.
- Evidence suggests that the Universe probably originated in a 'big bang'.

A Life cycle of a star

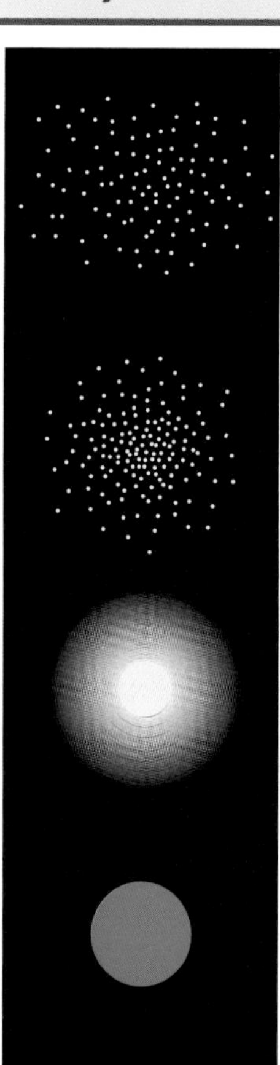

Stars are formed from massive clouds of dust and gas in space. **Gravity** pulls the dust and gas together. As the mass falls together it gets hot. **A star is formed** when it is hot enough for the hydrogen nuclei to **fuse together to make helium**. This **nuclear fusion process** releases **energy**, which keeps the **core** of the star **hot**.

During this stable phase in the life of the star, the force of gravity holding the star together is balanced by the high pressures caused by the high temperatures. **Our Sun** is at this **stable phase** in its life.

When all the hydrogen has been used up in the fusion process, **larger nuclei** begin to form and the star may,

- expand to become a **red giant**:
- or:

When all the nuclear reactions are over, a small star, like our Sun, may begin to contract under the pull of gravity. It becomes a **white dwarf**, which fades and changes colour as it cools down.

A larger, more massive, star may go on making nuclear reactions, getting hotter and expanding until it explodes as a **supernova**.

Some of the matter in the supernova is thrown off into space as dust and gas. Some of this dust and gas will go on to make new stars. Our Sun contains heavy elements, which suggests that it was probably created from the dust of an old supernova, along with the rest of the solar system, including your own body! – You are really 'star dust'.

What is left behind when the supernova explodes becomes a small dense **neutron star**, which shrinks, pulled in by its own gravity.

If the pull of gravity is strong enough, nothing can escape - not even light. The core of the former supernova has become a **black hole**.

Q Draw a flow chart to show the possible life routes of a star.

B The Milky Way and beyond

Remember
You should be able to describe the life of a star and the importance of gravity in the story.

1 The solar system is centred around our Sun, which is just one of many stars that make up the **galaxy called the Milky Way**. The stars in our galaxy are millions of times further apart than the planets in the solar system.

2 The Milky Way galaxy is just one of **many millions of galaxies in the Universe**. Some of the galaxies appear as points of light in the night sky. The galaxies are often millions of times further apart than the stars within a galaxy.

3 Observations of light from other galaxies show that their light appears to be shifted towards the red end of the spectrum. The further a galaxy is away, the greater this **red shift**.

KEY FACT

4 **An explanation for these observations is that all the galaxies in the Universe are moving apart very fast. The distant galaxies, with the bigger red shift, are moving away faster than those nearer to us. This suggests that the Universe was formed by a big bang, which threw all the matter out in different directions.**

5 The future of the Universe depends on exactly how much matter there is and how fast it is moving. **Gravitational** forces hold the Universe together. If there is **enough matter**, the force of gravitation may **slow down the expansion** and eventually **make the Universe contract**. If there is **less matter** the expansion may **go on forever**.

Q What does the red shift suggest about the Universe?

C Is there life out there?

1 With all these galaxies containing all these stars, is there another star out there with a planet that supports life – perhaps intelligent life?

2 Scientists are using the best telescopes they have to look for possibilities of life beyond Earth. So far, they have discovered over 70 planets orbiting stars other than our Sun. How might we detect if these or other planets support life?

3 The atmosphere of a planet supporting life might be very different from that of a dead planet - the Earth's atmosphere has far more oxygen than it would have if there were no living organisms.

4 Deep-space probes are travelling out to the edges of the solar system and beyond. They may even land robots on suitable planets to collect and analyse samples of material for signs of life.

Q What evidence would you want to see to be convinced that there is life on another planet somewhere in the Universe?

You might be asked to evaluate evidence for the possibility of life elsewhere in the Universe.

1 What is the likely end to our star, the Sun?

2 Why do people say that when you look at the stars you are looking back through history?

3 Light from the Sun takes a little over eight minutes to reach us. Light travels at 300 000 kilometres/second. How far away is the Sun?

Work, power and energy

- ➤ When a force makes something move, work is done and energy is transferred.
- ➤ An object gains gravitational potential energy when it is lifted up.
- ➤ Kinetic energy is the energy an object has because it is moving.

Remember
The pull of gravity, g, is 10 newtons on every kilogram
W = mg

Q A weightlifter lifts 100 kg through 1.5 metres. How much work does he do?

A Energy and work

1 Energy and work are closely related, and are both measured using the same unit – the joule.

- One kilojoule (kJ) = 1000 joules (J).

2 Work is what you do when you apply a force to move an object.

- When you lift a load you do some work - if you lift a heavy load, your arms will remind you that you are doing a lot of work. You are transferring energy from your muscles to the weight.
- When you push a broken-down car along the road you are doing work.

You have to exert a force to overcome the friction within the car engine and wheels and between the wheels and the ground. Food provides the energy for your muscles to do the work.

KEY FACT

3 Use this equation to calculate the work done when a force moves something through a distance:
work done (joules) = force (newtons) × distance (metres)

B Power

1 Power measures how quickly energy is transferred.

- Power is measured in watts (W).

KEY FACT

2 Calculate power from the equation:

$$\text{power (W)} = \frac{\text{work done or energy transferred (J)}}{\text{time taken (s)}}$$

- There is more about power on page 130.

3 **Example:** A boy with a mass of 50 kg runs up a flight of stairs in 5 seconds. The height he has risen is 20 m. How much power did he develop?

Answer: He uses a force of 500 N to lift his mass of 50 kg.

The work done in going up stairs = force × distance = 500 N × 20 m = 10000 J

$$\text{power} = \frac{\text{work done}}{\text{time taken}} = \frac{10\,000\text{ J}}{5\text{ s}} = 2000\text{ W}$$

Q Where does the energy come from to power the boy's muscles?

C Potential energy

Remember
The unit for energy is joule (J).

1 When the student puts the cans on the shelf she is doing **work**. Energy is transferred from the food in her muscles to the cans.

2 The cans **store energy** in the Earth's gravitational field. They have gained **gravitational potential energy**.

KEY FACT

3 The gain in gravitational potential energy depends on the height the cans are lifted and their weight.

$$\text{potential energy (J)} = \text{mass (kg)} \times \text{gravitational field strength (N/kg)} \times \text{height lifted (m)}$$

$$PE = mgh$$

4 **Example:** The student is filling shelves in the supermarket. Each can of beans weighs 0.5 kg and has to be lifted 1.2 m onto the shelf. How much potential energy does each can gain?

Answer:
Potential energy $= mgh$
$= 0.5 \text{ kg} \times 10 \text{ N/kg} \times 1.2 \text{ m}$
$= 6 \text{ J}$

Q How much potential energy is gained by a 70 kg person climbing 3 m?

You need to be able to recall and use potential energy = mgh.

D Kinetic energy

1 What happens to that potential energy when the cans fall? The cans fall to the ground, accelerating as they go.

- The moving cans have kinetic energy.

KEY FACT

2 The kinetic energy depends on the mass of the cans and how fast they fall.

$$\text{kinetic energy (J)} = \frac{1}{2} \times \text{mass (kg)} \times \text{speed}^2 \text{ (m/s)}^2$$

$$KE = \frac{1}{2}mv^2$$

You need to be able to recall and use kinetic energy = $\frac{1}{2}mv^2$.

PRACTICE

1 A builder carries 30 kg bricks up a ladder, lifting the bricks 5 m. How much potential energy does a brick gain?

2 A 70 kg man rides up a staircase using a stairlift. The lift carries him up a height of 3 m in 30 s.
a) Calculate the work done by the stair lift.
b) Calculate the useful power transferred by the stairlift.
c) The batteries driving the stairlift need to transfer more power than you calculated in (b). What else must the lift do?

Energy transfers

- ➤ Thermal energy can be transferred by conduction, convection or radiation.
- ➤ Evaporation also leads to the transfer of energy.

A Heat and temperature

Q How does the motion of the atoms in a gas change when the temperature rises?

1 Many energy transfers involve **heating**. Energy is contained in the **random vibrations and movements of atoms** in solids, liquids, and gases.

2 When an object becomes **hotter** (reaches a higher **temperature**) this means that the energy of motion of its individual atoms has **increased**.

3 The three main ways that heating transfers energy are **conduction**, **convection** and **radiation**.

- Energy can also be transferred by **evaporation**.

B Conduction

KEY FACT

1 **Energy is transferred from the hotter part of a solid to the colder part by conduction.**

- Energy is transferred by the particles in the material. The particles in the **hot** part are **vibrating more**. These vibrations are passed on to the cooler particles next to them, so the energy spreads through the material until all particles have the same energy.

2 Different materials vary as to **how well they conduct heat**:

Metals are good **conductors**, whereas most non-metals are poor conductors. Poor conductors are used as **insulators**. Most liquids and gases are poor conductors.

- The free **electrons** in metals have more kinetic energy when the metal is hot. The electrons help to transfer energy from the hot part of a metal to the cooler part.
- Insulators are used to keep things **cold** as well as keep things hot.

3 **Example:** Why do several layers of clothing keep you warmer than one thick layer?

Answer: The layers of clothing trap **air** between them, and **air is a good insulator**.

Q Why do electrons make metals good conductors?

C Convection

1 **Convection** is the transfer of energy by the movement of a liquid or gas.

2 When a liquid or gas becomes warm it expands and becomes less dense.
The warmer fluid floats above the cooler fluid, which sinks.
This creates a flow, which is called a convection current.

3 Most rooms are heated by convection currents. Warm air rises from a heater, and cooler air takes its place, to be heated in turn.

EY FACT

Q Why are the upper floors in a house often the warmest?

D Radiation

All bodies emit radiation. The hotter the body, the more energy it radiates. The radiation is usually in the **infrared** part of the electromagnetic spectrum.

- Dark, dull surfaces emit more radiation than light, shiny surfaces. They also absorb radiation well.
- Light, shiny surfaces do not absorb radiation well – they are good reflectors.
- Radiation can pass through space - that is how the warmth of the Sun reaches the Earth.

No energy transfer is 100% efficient - energy is often lost in warming up the surroundings.

Remember
Energy is transferred from anything that is hotter than its surroundings.

Q Can you suggest methods of cooking that use each of: conduction, convection and radiation?

E Evaporation

1 **Evaporation** is when the particles near the surface of a liquid **leave the liquid and become a vapour**.

2 The particles that escape are those with **higher-than-average energy** - otherwise they would not have enough energy to escape.

3 Because the liquid loses its higher energy particles, the liquid that is **left** has **less energy**, so it becomes **cooler**.

Q How does evaporation cause cooling?

PRACTICE

1 Suggest materials that could be used to insulate the roof space in a house.

2 Suggest why the central heating radiators in houses should perhaps be called 'convectors'.

3 Describe three ways in which energy is lost from the house and suggest ways for reducing the losses.

Energy resources

- Much of the energy that humans use comes from fossil fuels, such as coal and natural gas.
- Fossil fuels are non-renewable resources.
- Renewable sources of energy are those that are continually replaced – like the Sun, wind and waves.
- There are many ways of conserving energy in the home.

A Generating electricity – now

1 Electricity is a very convenient way of supplying energy.

2 Most UK power stations use fossil fuels - coal, oil and gas.

3 As well as producing electricity, fossil fuel power stations produce **waste gases** that contribute to **acid rain**, and **carbon dioxide**, a **greenhouse gas**. They also **warm up the surroundings**.

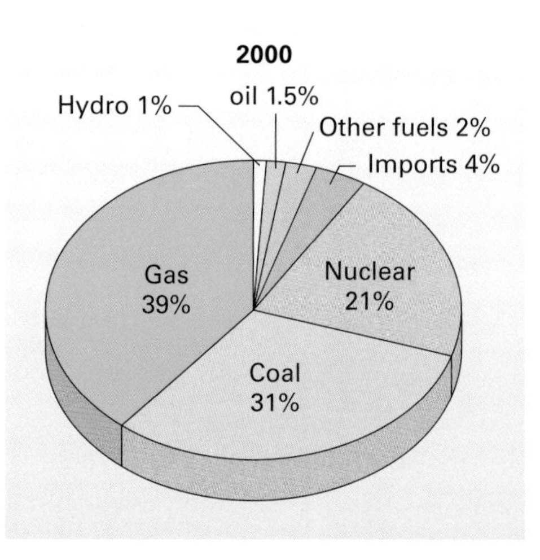

KEY FACT

4 Nuclear power stations do not produce gases that contribute to the greenhouse effect or that help to produce acid rain.

- Very little radiation escapes to the surroundings when the power station is running normally. However, the **waste** from the nuclear power station **remains radioactive for many years**.

5 **Example:** What percentage of Britain's electricity is generated using fossil fuels?

Answer: coal 31%; oil 1.5%; gas 39%; giving a total of **71.7%**.

Q We also use fossil fuels to power our cars. What has happened to the energy from the burning petrol by the end of the journey?

B Generating electricity – the future

1 Energy resources that do not get used up, or that are continually being replaced are called **renewable energy sources**.

2 Scientists and engineers are researching ways of using renewable energy sources to replace fossil fuels in power stations.

3 Wave power, wind power and solar panels **all depend on the weather**, so cannot be relied on alone. They need to be used with an **energy storage system** for back-up.

Q Think of four reasons why our society should try to reduce its use of electricity.

B

4 The pumped-storage hydroelectric power station at Dinorwig in Wales provides that back-up. At times of **low energy demand**, electricity is used to **pump water to the top reservoir**. When demand for electricity is **high** the **water flows back through the pipes** to drive the turbines and generate electricity.

C The green house

We could make more efficient use of energy and materials in our homes:

Solar water heating supplies half the hot water for the house.

Fit at least 25cm of insulation in the loft. Pay back time: two years.

Lagged pipes and hot water cylinder saves £20 per year.

Fit double- or triple- glazing with low e glass, that reflects heat back into the room.

Check walls - fit cavity wall insulation.

Turn down the thermostat to around 18–21°C. Put foil behind radiators.

Use energy-efficient light bulbs. Pay-back time 6 months.

Switch off the TV or computer when not in use.

Defrost the fridge regularly. Switch off the tumble drier. Hang clothes on a rack.

Remember
You need to be able to carry out calculations about efficiency.

Q Identify three places in the home where reducing heat losses can save money.

You need to be able to compare the advantages and disadvantages of using different types of energy sources.

PRACTICE

1 What else do we use fossil fuels for, apart from burning as a fuel?

2 A coal-fired power station has an efficiency of 35%. What does this mean?

3 We could use more public transport and private cars less. How would using public transport save energy?

4 What are the advantages and disadvantages of using wind or water to generate power?

What is radioactivity?

- ➤ Some isotopes of elements are radioactive – they give out radiation.
- ➤ The three kinds of radiation are alpha, beta and gamma.
- ➤ The different kinds of radiation travel at different speeds and penetrate objects to different degrees.
- ➤ Radiation ionises molecules in the material it passes through.

A Radioisotopes and radioactivity

Q What are the two types of particles that make up the nucleus of every atom?

1 The nucleus of an atom is made up of positively-charged **protons** and neutral **neutrons**. Each element has a **different number of protons**. (*For more on atoms see page 74.*)

2 Many elements have atoms that come in different versions, called **isotopes**.

- Although the isotopes of an element all have the **same atomic number Z** (that is, the same number of protons), they have **different numbers of neutrons**. **The mass number A** of different isotopes of an element will be **different**.

KEY FACT

3 The nuclei of some isotopes are unstable because they have too many neutrons. These isotopes emit radiation – they are radioactive and are called radioisotopes.

B Alpha (α) radiation

KEY FACT

1 Alpha radiation consists of a stream of alpha particles, each of which is the same as a helium nucleus.

Alpha radiation has:

- two protons and two neutrons
- two positive charges due to the two protons.

2 When an alpha particle is emitted from a radioisotope **the nucleus loses two protons and two neutrons** - its atomic number decreases by two, its mass number decreases by 4. **It has become a different element.**

$$^{224}_{88}Ra \rightarrow {}^{220}_{86}Ra + {}^{4}_{2}He$$

Q Why do alpha particles lose kinetic energy as they pass through the air?

3 Alpha particles ionise the air as they pass through it.

- Each time they ionise an air molecule they lose kinetic energy. They can only travel a few centimetres before losing all their kinetic energy.
- Alpha particles are stopped by paper.

C Beta (β) radiation

KEY FACT

1 Beta particles are high-speed electrons.

- Beta particles come from the nucleus of the atom, not from the cloud of electrons around the nucleus.

2 Inside the nucleus of a radioisotope, a neutron has decayed, to form a proton and an electron. A very small particle called an antineutrino is also produced. When a radioisotope emits a beta particle the mass number does not change, but the atomic number increases by one. It has become a different element.

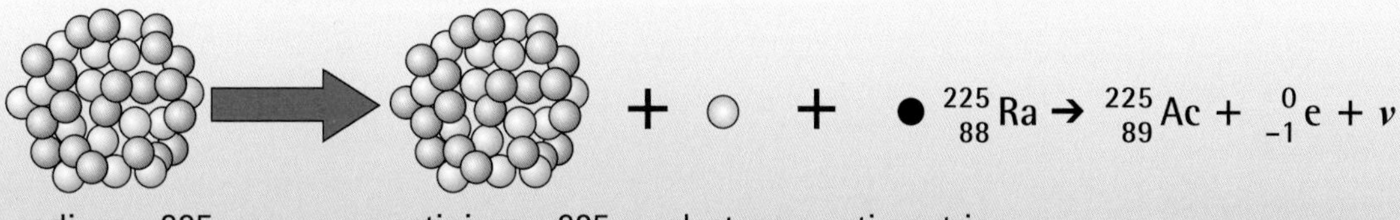

3 Beta particles are much smaller and lighter than alpha particles and move more quickly.

- they ionise air less well, and can travel a few metres before being stopped
- beta particles can pass through paper but are stopped by a few millimetres' thickness of aluminium.

Q What is the charge on a beta particle?

D Gamma (γ) radiation

KEY FACT

1 Gamma radiation is from the very short wavelength part of the electromagnetic spectrum.

2 Gamma radiation is emitted from an unstable nucleus when it loses energy - often after it has recently emitted an alpha or beta particle.

3 The mass number and atomic number do not change when gamma radiation is emitted.

4 Gamma radiation moves at the speed of light.

- gamma radiation is less ionising than alpha or beta radiation
- gamma radiation has very high energy and is only stopped by a very thick piece of lead or even thicker concrete.

Q What stops gamma radiation?

PRACTICE

1 Compete this table summarising the properties of alpha, beta and gamma radiations.

radiation	what is it?	how far can it travel in air?	what stops it?
alpha α			
beta β			
gamma γ			

Radioactive decay

- Background radiation comes from both natural and artificial sources in the environment.
- Radioactive decay is a random process.
- The half-life of a radioisotope is the time taken for half the nuclei present to decay.
- Radioactive decay processes can be used to date materials.

A Background radiation

KEY FACT

1 Background radiation is the <u>radiation that is all around us</u>. Most is <u>natural</u>, but some is caused by <u>human activity</u>.

- Some background radiation comes from **the Sun and outer space** - it is called **cosmic rays**. People who travel regularly in high-flying aeroplanes are exposed to more cosmic radiation than people on the ground.
- Some **rocks, such as granite**, emit radiation. Some parts of the UK e.g. Cornwall and parts of Derbyshire, have radioactive rocks, and buildings made from the rocks.
- Some **food** is radioactive - particularly shellfish.
- Radiation still lingers in the atmosphere and in the ground from **earlier nuclear events** - including the atomic bomb dropped at Hiroshima, Japan, in 1945 and the accident at the Chernobyl nuclear power station in the Ukraine in 1986.
- Some radiation in the environment is due to **waste** from **hospitals and nuclear power stations**.

2 Human bodies are used to these low doses of background radiation, but if the amount of radioactive material in the surroundings becomes **too high**, there could be **problems for your health**.

3 You have to take the **background radiation into account** when you are taking **radioactivity measurements**.

Q How could you reduce your exposure to background radiation?

B Half-life

1 Radioactive decay is a **random process** that helps the nucleus to become **more stable**. You **cannot predict** when a **particular nucleus** will decay, but you **can** say that after a certain amount of time, **half the nuclei in a sample** of the material will have decayed.

KEY FACT

2 The time it takes for half the nuclei in a sample to decay is called the half-life.

3 As the nuclei decay, there will be fewer left to decay later, so the **decay rate will decrease**. A graph showing how much radioactive material is present will show the rate at which the material decays:

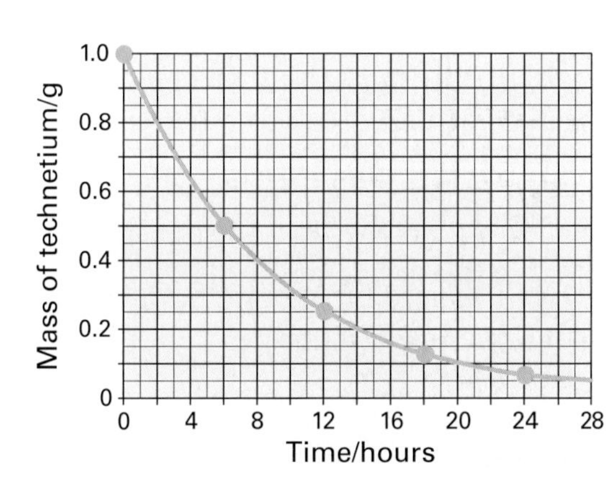

The graph shows how a sample of technetium decays. Technetium-99 has a half-life of six hours. This means that if there is one gram at the start, then after six hours this will have decayed to half a gram and another six hours later there will only be one quarter of a gram left.

Q What is half-life?

4 The half-lives of radioisotopes vary from **fractions of seconds** to **millions of years**.

C Dating using radioisotopes

Radioisotopes decay to make stable isotopes. Physicists can measure how much radioisotope is present in a sample of material and how much of the stable isotope is present. From this they can work out the age of the sample of material.

Carbon-14 is created from carbon-12 when cosmic rays bombard the atmosphere. Growing plants take in carbon-14 together with carbon-12 during photosynthesis. While the plant is alive the proportion of carbon-14 is constant. When the plant dies the carbon-14 decays, with a half-life of 5730 years. The technique of calculating age from these facts is called **radiocarbon dating**.

Q Why would radiocarbon dating be no use in trying to date something made from metal?

PRACTICE

1 A particular radioactive isotope has a half-life of two days. Twelve milligrams of the isotope are injected into the patient. After 6 days: (a) how many half-lives have passed? (b) How much of the isotope will remain in the patient?

2 The half life of radon-220 is 55 s. How long will it take for a sample of radon-220 to lose three-quarters of its radioactive atoms?

Using radioactivity

- Radiation can damage cells in the body – safety precautions are always needed when handling it.
- Radiation has a number of different applications in medicine and industry.

A Radiation safety

KEY FACT

1 Radiation can cause ionisation of the molecules in living cells. This damages the cells and may cause cancer.

2 Care must be taken when **handling radioactive materials**. People who work with ionising radiations must be particularly careful - the **greater the dose**, the greater the **risk of damage**. The key precautions are:

- Keep the radioactive source at a distance
- Use a suitable material to shield the user from the radiation
- Use the material for as short a time as possible
- Choose a suitable radioisotope for the job.

3 Radiation can be detected using a **Geiger-Müller tube**:

The tube contains argon gas, which **ionises** when a radioactive particle passes through it. A small current passes between the anode and the cathode. A counter detects this current.

4 People who work with ionising radiation wear a **film badge**. The badge contains photographic film that is sensitive to ionising radiation:

The badge is lined with different materials so that different parts of the badge detect different radiation.

5 **Example:** Why is it important for people working with radiation to monitor how much radiation they have been exposed to?

Answer: The more radiation a person receives, the higher chance there is that they may get cancer. There are **legal limits to the dose** that a worker should receive in any year.

Q What properties of radiation are used in its detection?

B Using radiation in medicine

Large doses of radiation are used to **kill cancer cells** in the body. Their use is carefully controlled. There is sometimes some **short-term damage to other healthy cells**. The different penetrating powers and ionisation properties are taken into account in choosing the radiation for the task.

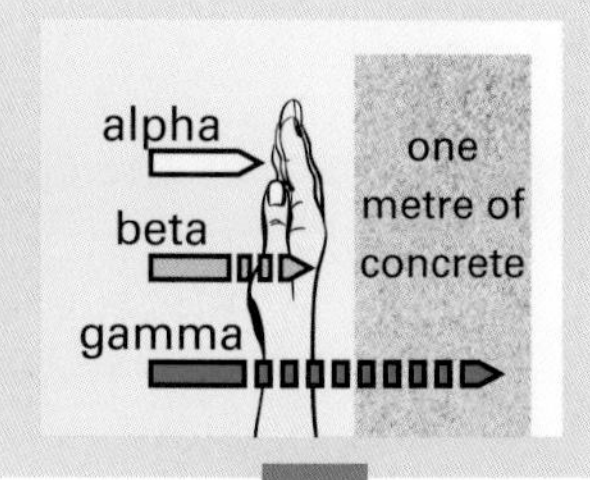

Gamma radiation can be used in similar ways to X-rays (see page 154). The radiation is directed from outside the body towards the cancer site.

The body would absorb **beta radiation** emitted from outside the body **before it reached** most cancers. However, iodine-131, a beta emitter, is used to treat **cancer of the thyroid gland** in the neck. The iodine is injected into the blood stream and taken up by the thyroid.

Alpha radiation is absorbed by air and would not penetrate the skin to treat a cancer, so it is **not suitable** for treatment of cancers from outside the body.

Radioactive isotopes are sometimes used as **tracers**. They can be used to find out if an organ in the human body is working properly. A radioactive material is injected into the bloodstream. The radiation that it emits can then be detected later to find its **position in the body**.

Q Which is the most penetrating radiation?

You need to be able to explain why a particular radiation is best for the job.

C Industrial uses of radiation

1 Radiation is **absorbed by the material** through which it passes. The **thicker** the material, the **greater the absorption**. This effect is used to **monitor thickness of materials** in production.

2 Some radioactive materials are used as **tracers** in industry. The material can be put into a fluid system, such as piped water. The radiation emitted **shows where the fluid has flowed**. This can be used to identify blockages or leakages.

- A short half-life radioisotope is used so it does not stay in the environment any longer than necessary for the tracing.
- A gamma ray emitter should be used, so that the radiation can be detected outside the pipe.

Q Why would gamma radiation be no use to detect paper thickness?

1 Explain the key precautions for radioactive safety.

2 Explain why alpha emitters are not used for the treatment of cancers.

3 Why is a gamma emitter best used for tracing fluid flows?

Biology

1 The graph shows the change in the total length of hedgerows in this country between 1940 and 1990.

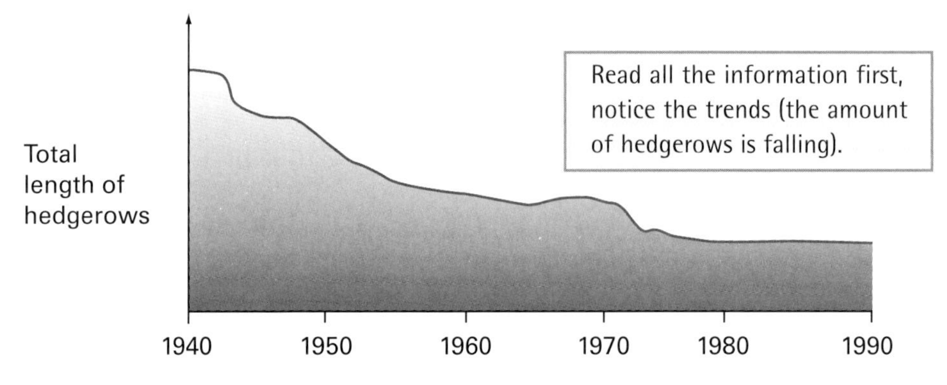

Read all the information first, notice the trends (the amount of hedgerows is falling).

Biodiversity refers to the number of different types of living organisms in a habitat. Explain how the change in the total length of hedgerows affects the biodiversity of this habitat.

- The amount of hedgerows decreases (almost halves in 30 years).
- Biodiversity (variety/numbers of types of species) drops.
- This is because hedgerow removal means loss of food, habitat, home, protection and breeding sites.
- Species become endangered/extinct.

These are the key words in the instructions. You need to state what happens, and explain why.

2 The graph shows the effect of the concentration of a plant hormone on the growth of roots and shoots.

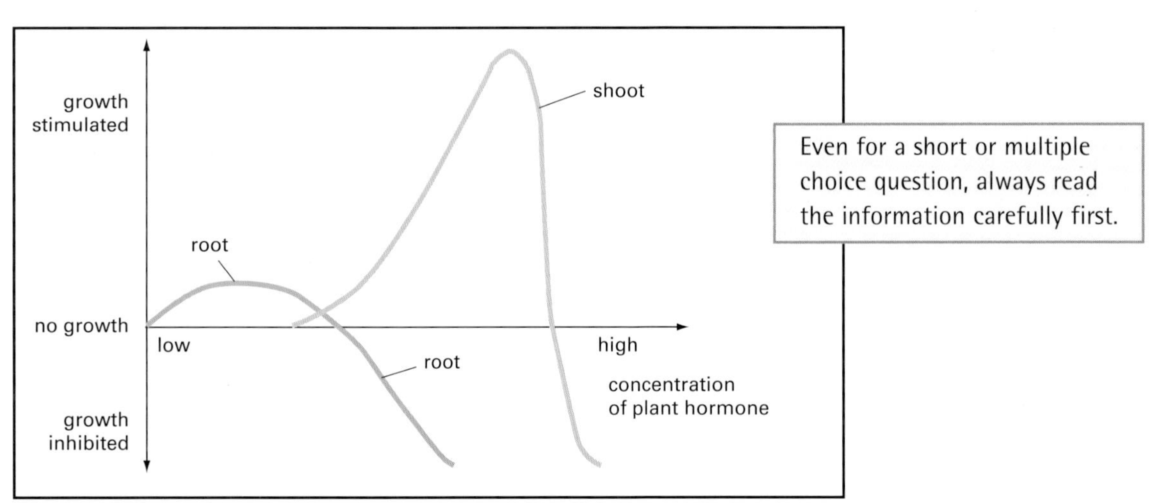

Even for a short or multiple choice question, always read the information carefully first.

Which statement is correct?

a) A change in the concentration of plant hormone affects the growth of shoots only.

This can't be true because both graphs show change.

b) Increasing the concentration of plant hormone always increases the growth of roots.

No, only at low concentrations.

c) A decrease in the concentration of plant hormone always slows down the growth of roots and shoots.

No, because higher concentrations slow growth.

d) A narrow range of concentration of plant hormone increases the growth of both roots and shoots.

Yes, although the range is different for roots and shoots.

Further questions for you to try

3 Dan measured the volume of urine he produced during the morning on two different days. His water intake was the same on both days. He did the same amount of exercise on both days. One day was hot, the other day was cold. His results are shown in the table:

	conditions	
	hot day	cold day
average volume of urine produced per hour (cm^3/h)	20	60

Explain how the volume of urine produced is affected by the hormone ADH.

4 Eric investigated the fermentation of solution X by yeast at different temperatures over a 24-hour period. He used the apparatus shown on the right:

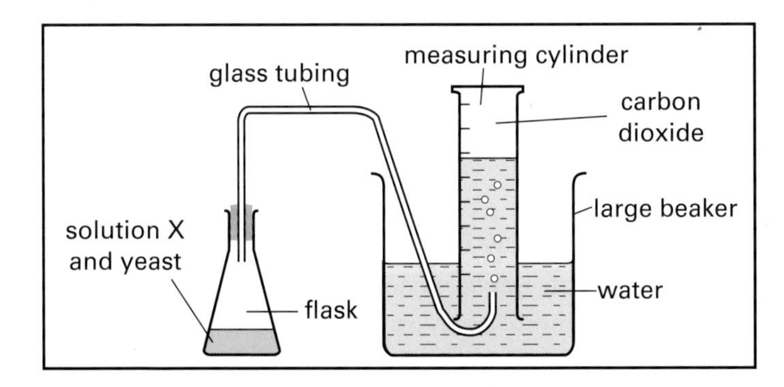

His results are shown in the table.

temperature (°C)	volume of carbon dioxide collected in 24 hrs (cm^3)			average volume of carbon dioxide collected in 24 hrs (cm^3)
10	25	28	28	27
20	58	65	66	63
30	99	108	102	_____

a) Calculate the average volume of carbon dioxide produced over 24 hours at 30°C. Show your working.

b) What is the advantage of calculating an average volume of carbon dioxide for each temperature?

c) What do Eric's results show about the relationship between temperature and the rate of fermentation?

d) What would be the effect of a temperature of 0°C on the rate of fermentation? GIve a reason for your answer.

5 Read the newspaper article about soya bean plants.

GM SOYA BEANS

Traditional soya bean plants are killed by selective weedkillers. Genetic engineers have transferred a gene into a soya bean plant to create a genetically modified variety. This new variety of soya bean p,ant is resistant to selective weedkiller. Farmers in the USA grow this variety and use selective weedkiller to improve crop yield.

Some people are concerned that there are dangers in growing genetically modified soya beans.

a) What is meant by the term **genetically modified**?

b) The new variety of soya bean plant is resistant to selective weedkiller. Explain how this can increase crop yield when the new variety is grown.

c) Suggest two ways in which growing the new variety of soya bean plants may be harmful.

6 The diagram shows a nephron.

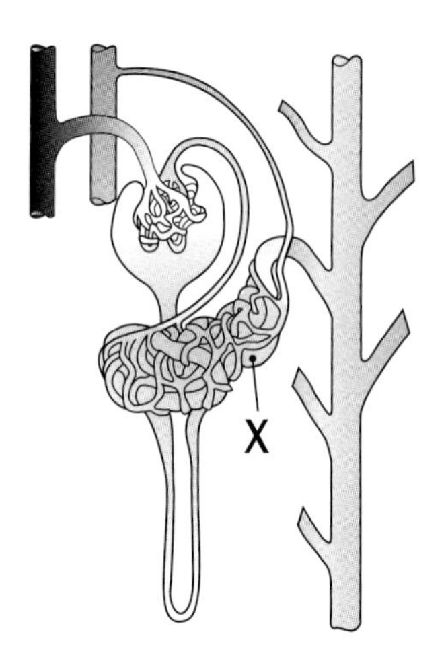

X is the:

a) first coiled tube

b) collecting duct

c) second coiled tube

d) Bowman's capsule?

Answers to further questions

3 Any three points from:
- less water in plasma/blood
- increased ADH / more ADH produced
- more water absorbed into blood
- urine volume decreases

4 a) (99 + 108 + 102) ÷ 3= 309 ÷ 3 = 103

b) more likely to be a true / realistic value

c) rate of fermentation increases between 10˚C and 30˚C

d) decreased rate/zero rate; enzyme/particles of reactants become less active

5 a) Organism that contains genes from another organism.

b) When soya bean is planted, weeds grow alongside. Weeds take space, block light and take nutrients from the soil. Killing the weeds means that crops plants are not deprived. Less light/nutrients = less growth.

6 c

Chemistry

1 Lithium is an alkali in group 1 of the Periodic Table.

a) Write the formula of
i) lithium chloride ii) lithium oxide iii) lithium carbonate.

i) $LiCl$ ii) Li_2O iii) Li_2CO_3

b) Explain, in terms of electron transfer, how lithium forms compounds.

Group 1 elements have one outer electron which is lost in reaction forming unipositive ions, i.e. Li – e^- ➔ Li^+, which are held to be negative ions in ionic compounds.
2.1 2

c) Name the process by which lithium would be obtained from its compounds and say why this method is chosen.

Electrolysis because lithium is a reaction metal (as are all the group 1 metals) with stable compounds, which cannot be reduced back to lithium by chemical methods.

d) Lithium reacts quickly with cold water.
i) Write a balanced equation for this reaction and name the products.
ii) What effect would the solution remaining have on universal indicator. What would this tell you?

i) $2Li(s) + 2H_2O(l) \rightarrow 2LiOH(aq) + H_2(g)$
lithium hydroxide, hydrogen

ii) the universal indicator would be purple; lithium hydroxide is a strong alkali

e) Name two metals that react more violently with water than lithium.
(Note to students: it is unlikely that you will have studied or have notes on lithium particularly. The question requires an understanding of the chemistry of the group 1 alkali metals and then its application to the chemistry of lithium. The most important thing to realise is that group 1 metals form unipositive ions, i.e. they have a valency/combining power of 1.)

You will probably not have seen lithium in the reactivity series as it is not usually one of the examples given; for this question you need to remember that reactivity increases as you go down group 1; hence any two metals below lithium in the group will be correct. The obvious answer is therefore sodium, potassium.

2 The energy level diagram for the burning of methane in oxygen is shown below:

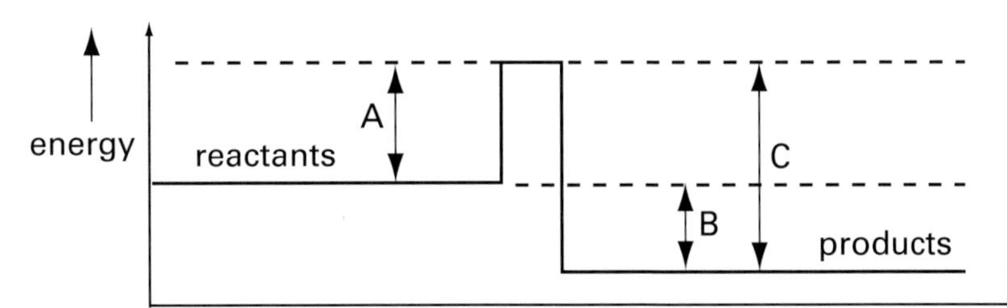

The energy level diagram on the previous page may be presented in a way that is different to the way you are used to. Do not be put off by this – the principles are exactly the same.

The question requires little or no factual knowledge; rather, a thorough understanding of the principles of energy transfers in reactions is what is needed.

a) Write a balanced equation for the reaction.

$CH_4(g) + 2O_2(g) \rightarrow CO_2(g) + 2H_2O(g)$

Note: combustion of hydrocarbons always produces CO_2 / H_2O.

b) Explain A, B, C on the diagram in terms of energy transfers occurring during the reaction.

A = activation energy for the reaction, energy put in to break reactant bonds
B = heat of reaction, ΔH
C = energy given out when product bonds are formed

c) Which bonds are broken and which formed during this reaction?

Write out the structural equation - in rough or in your answer - to show you what bonds are present.

i.e. H–C(H)(H)–H + O = O, O = O → O = C = O + H – O – H, H – O – H

bonds broken: 4 × C–H and 2 × O = O
bonds formed: 2 × C = O and 4 × O–H

d) Is the reaction exothermic or endothermic?

the reaction is exothermic

You know this because products are at a lower energy level than reactants.

3 Ammonia is manufactured from nitrogen and hydrogen in the Haber process, the equation for which is:

$N_2 + 3H_2 \rightleftharpoons 2NH_3$; ΔH is negative

a) What does the symbol $\rightleftharpoons$ mean?

The symbol $\rightleftharpoons$ means the reaction is reversible and the products can react to give the original reactants back again.

b) This reaction requires equilibrium; what does this mean?

Equilibrium is the point at which rate forward = rate backward and the amounts of substances present are not changing.

c) What else is present in the equilibrium mixture along with ammonia?

Nitrogen and hydrogen are also present.

d) How is the amount of ammonia present at equilibrium affected by:
i) an increase in temperature ii) an increase in pressure?

i) If the temperature is raised the system will move to absorb or minimise this rise (Le Chatelier's principle) and will move to the left.
ii) If the pressure is increased the system will try to absorb the pressure and move to where the number of molecules is least. The system will move to the right.

e) How is the rate at which equilibrium is reached affected by temperature?

If the temperature is increased the molecules will gain energy and thus the number of fruitful collisions possessing activation energy for the reaction will increase; hence rate increases with temperature increase.

f) Explain why iron is used in this process.

Iron is used as a catalyst; it speeds up the attainment of equilibrium but has no effect on position.

Questions like this on reversible reactions/equilibrium etc. are very common and usually involve the Haber process. You should make sure you have learned the facts of the process and understand the principles of changing the position of equilibrium, etc. thoroughly. Be ready for different examples where the smallest side might be the left and ΔH may be positive. The principles are exactly the same but the equilibrium movements are the opposite of those you might be used to in the examples you have seen.

Further questions for you to try

4 a) Describe (with balanced equations) how magnesium and zinc react with dilute hydrochloric acid.
b) Predict how i) calcium ii) gold will react with dilute hydrochloric acid, writing balanced equations.
c) What would you expect to see if a few pieces of zinc were left in some blue copper (II) sulphate solution? Write a balanced equation for the reaction is appropriate.

5 Aluminium is extracted from its ore by an electrolytic process.
   a) Name the ore and the compound of aluminium it contains.
   b) Explain how this ore is made liquid for the electrolytic process and why this method is used.
   c) Explain, with balanced equations, the chemistry of the process.

6 If dilute hydrochloric acid is added to marble chips (calcium carbonate) the reaction proceeds easily.
   a) Write a balanced equation for the reaction.
   b) State three ways in which the reaction could be speeded up and in each case explain clearly why the action taken results in an increase in rate.

7 a) How did the original hydrogen and helium disappear from the atmosphere?
   b) How did the gases present 3500 millions years ago arrive?
   c) What caused the decrease in CO_2 levels at about 300 m. years ago?
   d) What caused the build up of oxygen?
   e) What effect did the formation of an ozone layer have?
   f) Give two reasons why the CO_2 concentration in the atmosphere has been gradually building up over the past 50 years?

Answers to further questions

4 a) Magnesium reacts quickly – bubbles of hydrogen seen in quantity.
$Mg(s) + 2HCl(aq) \rightarrow MgCl_2(aq) + H_2(g)$.
Zinc reacts easily but less readily – a slow stream of hydrogen.
$Zn(s) + 2HCl(aq) \rightarrow ZnCl_2(aq) + H_2(g)$

b) i) Calcium is a reactive metal and the reaction is rapid and dangerous.
$Ca(s) + 2HCl(aq) \rightarrow CaCl_2(aq) + H_2(g)$

ii) Gold is very unreactive and no reaction takes place.

c) The zinc would displace the copper and gradually become coated with red-brown copper powder; the blue of the copper (II) sulphate solution becomes paler until eventually it is colourless zinc sulphate solution.
$Zn(s) + CuSO_4(aq) \rightarrow ZnSO_4(aq) + Cu(s)$

5 a) bauxite, aluminium oxide (Al_2O_3)

b) The aluminium oxide is dissolved in molten cryolite, a compound of aluminium; aluminium oxide has a melting point of over 2000°C and so hot, molten aluminium oxide is at too high a temperature; cryolite melts well below 1000°C and this is a safer temperature at which to perform the electrolysis.

c) ions present: $Al_2O_3 \rightarrow 2Al^{3+} + 3O^{2-}$
at anode (+): $3O^{2-} - 6e^- \rightarrow 3O$
$3O + 3O \rightarrow 3O_2(g)$
at cathode (−): $2Al^{3+} + 6e^- \rightarrow 2Al(s)$
overall: $2Al_2O_3 \rightarrow 4Al(s) + 3O_2(g)$

6 a) $CaCO_3(s) + 2HCl(aq) \rightarrow CaCl_2(aq) + H_2O(l) + CO_2(g)$

b) i) Increase the temperature. This gives energy to the particles thus increasing the number of collisions with the necessary activation energy.

ii) Increase surface area of marble chips by having smaller pieces or powdered chip. This increases the number of $CaCO_3$ particles available to the HCl particles and thus increases the likelihood of fruitful collisions.

iii) Increase the concentration of the hydrochloric acid. This increases the number of HCl particles and thus increases the likelihood of fruitful collisions.

Note: This reaction is generally not catalysed and adding a catalyst is therefore not an acceptable answer.

7 a) Light gases, which floated off into space.
b) Emitted from volcanic activity.
c) Plants began to appear and to absorb CO_2 in photosynthesis.
d) More and more plants on the earth's surface were producing more and more oxygen by photosynthesis.
e) Filtered harmful UV light from the sun thus allowing more complex animal life to exist.
f) Cutting down rain forests thus removing plants which absorb CO_2; burning more fossil fuels thus putting more CO_2 into the atmosphere.

1 A thundercloud forms above a church spire.

Negative charge has collected at the bottom of the cloud. In a lightning discharge, 15 coulombs of charge transfers from the cloud to the Earth. The discharge takes 0.0005 seconds and produces a very large current.

a) Calculate the value of this current.

- Write down the equation.
- Write in the values you know.
- Write out the answer with the units.

current = 30 000 A

b) The energy transferred from the cloud to the Earth during the discharge is 150 MJ. Calculate the voltage (in MV) between the cloud and the Earth that transfers the 15 coulombs of charge.

- Write down the equation.
- Write in the values you know.
- Write out the answer with the units.

voltage = 10 MV

c) A lightning conductor is connected to the spire to prevent damage. It has a metal spike on the spire connected by a thick copper cable to a metal bar in the ground (earth). Why is the copper cable thick?

A lot of energy is transferred to the ground very quickly. This will heat up the copper cable, which might melt if it was not thick enough.

Think about the energy and current values you have been working with.

2 A hovercraft is at rest on land.

a) The hovercraft travels between Portsmouth and the Isle of Wight. It takes half an hour (0.5 hour) to make the 15 km journey. Calculate the average speed of the journey.

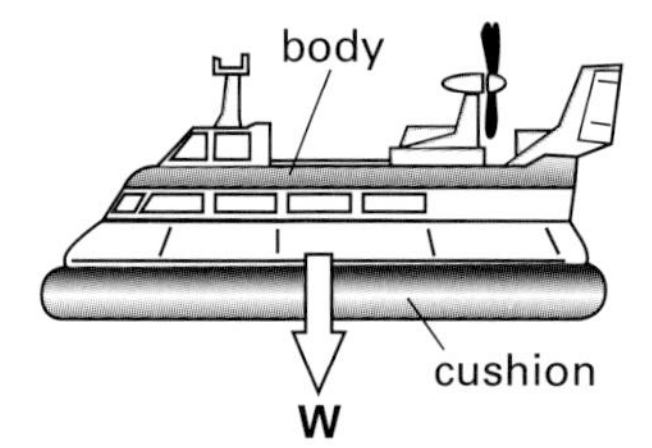

- Write down the equation.
- Write in the values you know.
- Write out the answer with the units.

speed = 30 km/h

b) The mass of the hovercraft is 40 000 kg. The unbalanced force on it is 20 000 N. Calculate the acceleration.

- Write down the equation. F = ma
- Write in the values you know. N = 40000 kg × a
- Rearrange the equation.
- Write out the answer with the units.

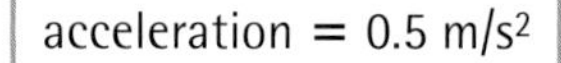

c) After a while the hovercraft reaches a steady speed even though the forces pushing it forwards stay the same. Explain.

If the hovercraft is moving at a steady speed the forces on it must be balanced. When the hovercraft accelerates it gets faster. As it gets faster the air resistance will get faster, until the air resistance equals the driving force – then it will move at a steady speed.

Think about the other forces acting on the hovercraft as well as the driving force.

Further questions for you to try

3 a) The diagram shows part of an optical fibre. Describe and explain the path of the electromagnetic wave as it passes along the fibre.

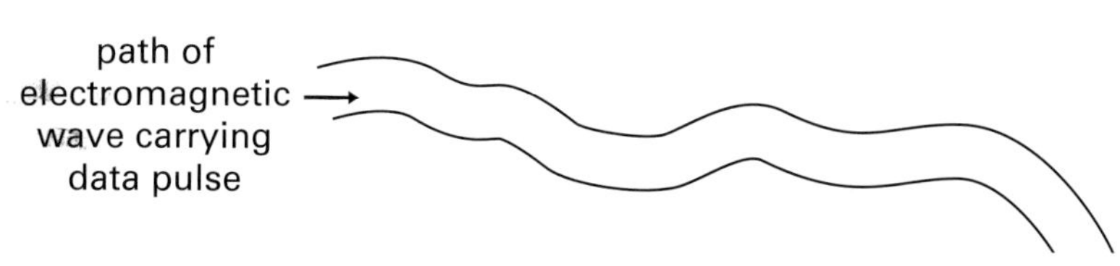

b) Doctors use endoscopes to see inside a patient's stomach. The diagram shows part of the endoscope. It shows two bundles of optical fibres inside a plastic tube.

i) Explain why endoscopes must have two bundles of optical fibres.

ii) Explain why one of the bundles of fibres must be arranged in the same pattern at both ends.

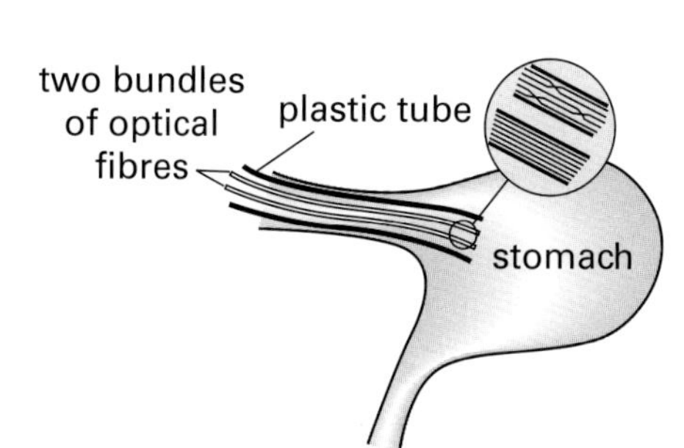

4 The light from stars from distant galaxies are said to red-shifted. Measurements of red shift allow astronomers to calculate the speed that galaxies are moving away from us.

a) What is meant by red shift?

b) What type of shift would be observed in the light from stars travelling towards us?

5 a) Energy can be transferred by conduction, convection and radiation. The diagram shows a solar panel used to heat water. Explain why the water coming out is a lot warmer than the water going in. Use your ideas about energy transfer.

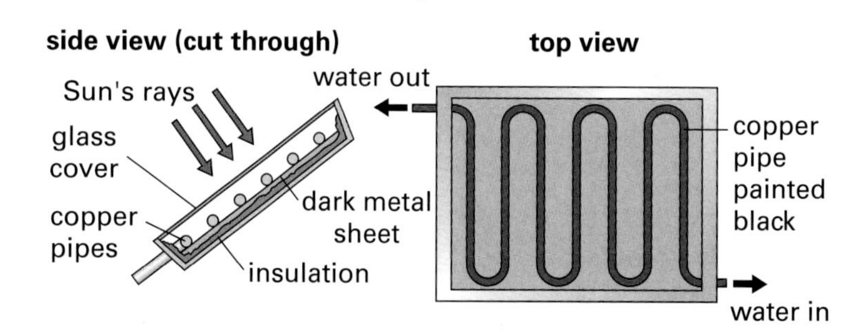

b) Explain how energy is transferred through a brick wall by conduction. Use your ideas about particles.

c) Materials that are good electrical conductors are also good at transferring energy by conduction. Explain why.

Answers to further questions

3 a) The waves strike the insides of the fibre at angles greater than the critical angle, so they are reflected back inside the fibre.

b) i) One bundle carries the light down to illuminate inside the stomach. The other bundle carries the reflected light back to form an image in a camera or for the doctor to see.

ii) The bundle carrying the image must have the same pattern, so the image is not scrambled.

4 a) The light emitted from the galaxy has a longer wavelength than expected. It has shifted to the red end of the spectrum.

b) blue shift

5 a) Infrared radiation from the Sun passes through the glass and is absorbed by the black pipes. The energy is conducted through the copper pipe. The water is warmed. Warm water is less dense than cold water so a convection current carries the warm water to the top of the panel where it leaves the panel, to be replaced by colder water at the bottom.

b) The particles in the brick vibrate. The energy is passed from atom to atom through the brick.

c) A conductor of electricity has many free electrons. These are free to move within the solid and transfer the energy through the material.

Topic checker

- Go through these questions after you've revised a group of topics, putting a tick if you know the answer, a cross if you don't – you can check your answers on the page references given.
- Try these questions again the next time you revise... until you've got a column that's all ticks! Then you'll know you can be confident...

Life processes

1 What are the seven main life processes?	☐	☐	☐
2 What are the main types of digestive enzymes?	☐	☐	☐
3 Why are vitamins and minerals important for nutrition?	☐	☐	☐
4 Why is the shape of an enzyme molecule important for its function?	☐	☐	☐
5 What are 5 differences between arteries and veins?	☐	☐	☐
6 What are the main components of blood?	☐	☐	☐
7 What is meant by immune response, and how can it be enhanced by medical science?	☐	☐	☐
8 Why is air pressure important in breathing?	☐	☐	☐
9 What word equations are used to represent anaerobic respiration a) in yeast cells b) in humans?	☐	☐	☐
10 What are the main components of the nervous system?	☐	☐	☐
11 How does alcohol affect the nervous system?	☐	☐	☐
12 What reflex action does the iris carry out?	☐	☐	☐
13 How do insulin and glucagon help to control blood sugar level?	☐	☐	☐

Answers 1 respiration; feeding/nutrition; sensitivity; movement; reproduction; growth; excretion 2 carbohydrase (break down carbohydrates); lipase (break down fats/lipids); protease (break down proteins) 3 for general health; some are needed for enzymes to work properly 4 active site of an enzyme molecule has a unique shape where particles can 'lock on' and react 5 arteries: more muscular walls, no valves, smaller central canal/lumen, carry oxygenated blood, carry blood away from heart towards organs (valves opposite to this); speed of blood flow fast 6 cells: red and white; platelets/cell fragments; plasma containing dissolved substances and plasma proteins 7 production of antibodies by white cells due to the presence of antigens/foreign proteins; enhanced by immunisation 8 difference in pressure/ pressure gradient between the inside of the lungs and the outside of the body causes inhalation/exhalation 9 yeast: glucose → ethanol/alcohol + carbon dioxide; human: glucose → lactate + carbon dioxide 10 receptors/sensory nerves or sense organs; central processing unit/the brain and spinal cord; motor neurones 11 sedative, slows response time, affects behaviour and judgement 12 automatically adjusts to amount of light entering eye 13 insulin lowers blood sugar level, mainly by causing the conversion of glucose to glycogen; glucagon is the opposite

14 Which body changes occur at puberty?	☐ ☐ ☐
15 How is the structure of the uterus suited to its function?	☐ ☐ ☐
16 How can fertility be altered by hormones?	☐ ☐ ☐
17 Why is it important to control body conditions?	☐ ☐ ☐
18 Which main stages of water and salt regulation occur in a) the Bowman's capsule b) the first part of the tubule c) the second part of the tubule?	☐ ☐ ☐

Plants

19 What are the word and symbol equations for photosynthesis?	☐ ☐ ☐
20 How is glucose from photosynthesis assimilated into plant tissues?	☐ ☐ ☐
21 What is meant by a concentration gradient in osmosis?	☐ ☐ ☐
22 Why is a waxy cuticle found on the upper leaf surface while stomata are mostly on the lower surface?	☐ ☐ ☐
23 How are auxins used commercially?	☐ ☐ ☐

Variation and inheritance

24 What are the main reasons for variation between individuals?	☐ ☐ ☐
25 What were the main points of Charles Darwin's theory of evolution by natural selection?	☐ ☐ ☐
26 What are two main differences between mitosis and meiosis?	☐ ☐ ☐
27 How is cloning used in agriculture?	☐ ☐ ☐

Answers 14 growth spurt; pubic hair growth; production of sex cells/eggs/sperms; penis and muscle development in males; breast development and widening of hips in females
15 can expand to a much bigger size, muscular to deliver the baby during birth; protected from the surroundings; warm
16 FSH used to boost egg production; oestrogen/progesterone in contraceptive pill used to block egg production
17 enzymes can be destroyed by change in cell conditions, hence cells stop functioning 18 a) ultrafiltration/materials filtered into capsule from blood b) glucose, water and other materials reabsorbed from tubule back into blood
c) fine tuning of water and salt balance
19 carbon dioxide + water → glucose + oxygen
$6CO_2 + 6H_2O \rightarrow C_6H_{12}O_6 + 6O_2$
20 glucose converted to starch for storage 21 a difference in the concentration of water molecules between one area or one cell, and another 22 upper surface generally in more direct sunlight so more likely to dry out, and waxy cuticle protects against water loss; stomata allow water loss, so their position on underside of leaf ensures that the rate of loss is minimal 23 stimulate root growth in stem cuttings; regulate fruit ripening; cause abnormal growth and plant death in weedkillers 24 genetic due to the inheritance of different DNA/genes/chromosomes; environmental due to factors in the environment/lifestyle 25 genetic variation gives rise to new characteristics; some characteristics are beneficial for survival; better adapted individuals are more likely to reproduce and pass on beneficial characteristics 26 mitosis: involved in asexual reproduction; all new cells are identical genetically and identical to parents; occurs during growth: meiosis: involved in sexual reproduction; there is genetic variation between new cells and parents; occurs during formation of sex cells/gametes 27 used to produce many more individuals the same as the original e.g. splitting embryos in cattle; growing cauliflowers from tissue culture

28 What is meant by a) phenotype and b) genotype?

29 What is biodiversity and why is it important?

30 How are the population sizes of a predator and its prey linked?

31 Which processes add carbon dioxide to the atmosphere and which remove it?

32 How is a pyramid of biomass drawn?

Atomic structure

33 How many protons, neutrons and electrons does $^{23}_{11}Na$ contain?

34 Chlorine gas contains 75% $^{35}_{17}Cl$ and 25% $^{35}_{17}Cl$.
Calculate its relative atomic mass.

35 What is the electronic configuration of the unknown element $_{22}X$?

Bonding

36 What is an ionic bond and why do ionic compounds have high melting points?

37 What is the charge on the ions formed from a) $_{20}Ca$ b) $_{9}F$?

38 Sketch a dot and cross diagram of the methane molecule.

39 Why do covalent compounds have low boiling points?

Chemical reactions

40 Write balanced chemical and ionic equations for:
magnesium + aluminium oxide → magnesium oxide + aluminium.

41 How many molecules are there in 64 g of oxygen gas (O = 16)?

42 What mass of hydrogen is needed to produce 9 g of water in:
$2H_2 + O_2 \rightarrow 2H_2O$ (H = 1, O = 16)?

Answers 28 phenotype: how characteristics are expressed e.g. hair colour: genotype: the types of alleles governing a characteristic 29 different groups of people have different interests with respect to the Earth's resources e.g. whether to conserve wild areas of land or use for agriculture 30 predator population will follow that of the prey with a time-lag 31 add carbon dioxide: respiration, burning/combustion of carbon based compounds, decay/rotting; remove carbon dioxide: photosynthesis 32 using a set unit on the horizontal scale for a unit of mass e.g. 1 cm = 1 kg 33 11p 12n 11e 34 $\frac{(75 \times 35) + (25 \times 37)}{100} = 35.5$ 35 2.8.10.2 36 the electrostatic attraction between positive and negative ions; these ions are in a lattice structure which is difficult to break down by heat 37 Ca^{2+}, F^{-}
38

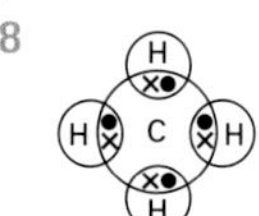

39 The forces between the separate molecules are very low hence the molecules are easily separated one from another.
40 $3Mg + Al_2O_3 \rightarrow 2Al + 3MgO$
$3Mg + 2Al^{3+} + 3O^{2-} \rightarrow 2Al + 3Mg^{2+} + 3O^{2-}$
or $3Mg + 2Al^{3+} \rightarrow 2Al + 3Mg^{2+}$
41 $2 \times 6.02 \times 10^{23}$ 42 1g

43 If gram molar volume is $24dm^3$ at r.t.p., what is the volume of 14g of nitrogen at r.t.p. (N = 14)?

44 Write the simplest ionic equation for neutralisation.

Organic chemistry

45 State the molecular and structural formula of propane.

46 What is the difference between saturated and unsaturated hydrocarbons?

47 What type of reaction is $CH_4(g) + Cl_2(g) \rightarrow CH_3Cl(l) + HCl(g)$?

The Earth and the atmosphere

48 Explain why zinc can be extracted by reduction of its ore but sodium is extracted by electrolysis.

49 Write the anode and cathode half-reaction equations for the electrolysis of potassium chloride (K^+Cl^-).

50 State the equation and conditions for converting nitrogen and hydrogen to ammonia.

51 What advantages do artificial fertilisers have over natural fertilisers?

52 What part did the ozone layer play in the evolution of life on earth?

The elements

53 What do the elements in a group of the periodic table have in common?

54 Element X is an alkali metal. Write a balanced equation for its reaction with water.

55 Write the two half-reaction electron-transfer equations for:
$Cl_2(g) + 2KI(aq) \rightarrow 2KCl(aq) + I_2(s)$

Measuring reactions

56 What is meant by the activation energy of a reaction?

57 How do catalysts increase the speed of reactions?

Answers 43 $12dm^3$ 44 $H^+ + OH^- \rightarrow H_2O$ 45 C_3H_8

```
  H H H
  | | |
H-C-C-C-H
  | | |
  H H H
```

46 saturated hydrocarbons contain only single bonds; unsaturated hydrocarbons contain at least one double bond 47 substitution 48 sodium is a very reactive metal with stable compounds that can only be decomposed by electrolysis; zinc is less reactive so its compounds are less stable and can be chemically reduced 49 anode (+): $2Cl^- - 2e^- \rightarrow Cl_2(g)$; cathode (–): $K^+ + e^- \rightarrow K(s)$ 50 $3H_2 + N_2 \xrightarrow[\text{Fe catalyst}]{250atm/550°C} 2NH_3$ 51 high yields of large, healthy plants quickly 52 filtered harmful UV radiation from the sun out; enabled complex animal life to develop 53 all have same number of outer electrons; similar chemical properties 54 $2X + 2H_2O \rightarrow 2XOH + H_2$ 55 $Cl_2 + 2e^- \rightarrow 2Cl^-$; $2I^- - 2e^- \rightarrow I_2$ 56 energy the molecules must have to produce fruitful collisions 57 reduce the activation energy for the reaction

58 What happens to the equilibrium A + B $\rightleftharpoons$ C + D, ΔH is positive, if the temperature is raised.

Electricity and magnetism

59 What is the relationship between current I, charge Q and time t?

60 What is the relationship between voltage V, current I and resistance R?

61 How can you change the direction of the force on a current-carrying coil of wire in a magnetic field?

62 Why do transformers need alternating current?

Force and motion

63 What are the factors that affect the stopping distance of a car?

64 What is the difference between mass of an object and its weight, and what are their units?

Waves

65 What is the difference between transverse and longitudinal waves – give an example of each.

66 What is the relationship between the frequency f, the speed v and the wavelength ? of a wave?

67 What is refraction of waves?

68 What is ultrasound and how is it used?

69 How are the differences between P earthquake waves and S earthquake waves used to find out more about the structure of the Earth?

Answers 58 moves to right to absorb the rise in temperature 59 charge = current × time; Q = I t 60 V = I R 61 Change the direction of the current or the direction of the magnetic field. 62 A changing current in the primary produces a changing magnetic field in the core of the transformer, which induces a changing voltage in the secondary coil. 63 The stopping distance depends on the thinking distance and the braking distance. 64 Mass of an object is measured in kilograms; weight of an object is a force incorporating gravity and is measured in newtons 65 In transverse waves the oscillation is at right angles to the direction the wave travels – e.g. light, water waves. In longitudinal eaves the oscillation is in the same direction that the wave travels – e.g. sound waves. 66 $v = f \lambda$ 67 Refraction occurs when a wave passes from one medium to another, the speed changes and the wave changes direction e.g. when light passes from air to glass. 68 Ultrasound is sound of frequencies too high to be heard. Used for imaging and also to measure distances. 69 P waves are longitudinal waves; S waves are transverse waves and can only travel through solids, so cannot pass through the core of the Earth. P waves travel more quickly than S waves. After an earthquake seismometers detect the time it takes for the waves to be felt around the Earth, this gives information about the exact location of the earthquake and the materials through which the waves passed.

70 What are the three main effects on the surface of the Earth of the movement of the tectonic plates?	☐ ☐ ☐

Earth and beyond

71 How do comets differ from planets?	☐ ☐ ☐
72 How are stars formed?	☐ ☐ ☐
73 What do observations of red shift tell us about the Universe?	☐ ☐ ☐

Energy

74 What are the relationships between energy, force, work, distance, power and time?	☐ ☐ ☐
75 What are the units used to measure energy, force, work, distance, power and time?	☐ ☐ ☐
76 What equations are used to calculate kinetic energy and change in gravitational potential energy?	☐ ☐ ☐
77 Explain in terms of particles how energy is transferred by conduction through a metal.	☐ ☐ ☐

Radioactivity

78 What are three components of an atom?	☐ ☐ ☐
79 What are the properties of alpha particles?	☐ ☐ ☐
80 What are the properties of beta particles?	☐ ☐ ☐
81 What are the properties of gamma rays?	☐ ☐ ☐
82 What is meant by the half-life of a radioactive element?	☐ ☐ ☐

Answers 70 Where the plates move apart hot molten rock rises and solidifies to form new plate material. Where the plates move together mountains may form by folding, volcanoes may erupt and earthquakes occur. Where plates slide past each other earthquakes occur. 71 Comets are lumps of ice, dust and gas; their orbits are very elliptical. They pass close to the Sun, when they can be seen, and then travel far beyond Pluto. 72 A cloud of dust and gases is drawn together by gravity, when the forces between the materials are large enough and the material gets hot enough nuclear fusion takes place and light is given out. 73 All the galaxies in the Universe are moving apart very fast. The distant galaxies, with the bigger red shift, are moving away faster than those nearer to us. This suggests that the Universe was formed by a big bang, which threw all the matter out in different directions. 74 work done = energy transferred; work = force × distance moved in the direction of the force 75 Energy, J; force, N; work, J; distance, m; power, W; time, s 76 change in gravitational potential energy = m g h; kinetic energy = $\frac{1}{2}m\,v^2$ 77 The particles in the hot part are vibrating more. These vibrations are passed on to the cooler particles next to them, so the energy spreads through the material until all particles have the same energy. The free electrons in metals have more kinetic energy when the metal is hot. The electrons help to transfer energy from then hot part of a metal to the cooler part. 78 electrons, protons and neutrons 79 Alpha particles are made of 2 protons and 2 neutrons; they are positively charge and stopped by paper. 80 Beta particles are electrons and are negatively charged, they are stopped by a few millimetres of aluminium. 81 Gamma rays are electromagnetic waves; they are stopped by several centimetres of lead or a greater thickness of concrete. 82 The half-life of a radioactive element is the time it takes for half the radioactive nuclei in a sample of the material to decay.

Complete the facts

1 Complete the labels for the diagram and phrases in the boxes.

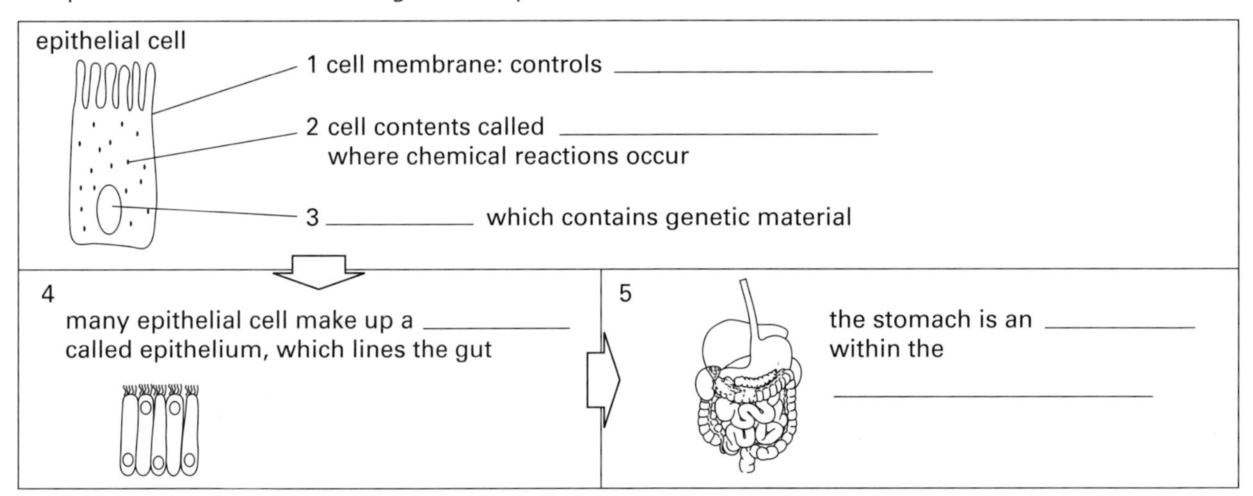

2 Complete the flow chart for the digestion of egg on toast:

main food types:
egg = ________
butter = ________
bread = ________

In the mouth		In the stomach		In the small intestines
__________		__________		________ remaining
is broken down	➔	is broken down	➔	food is digested, and
by ________		by ________		________ into the
into ________.		into ________.		bloodstream.

3 Complete the notes on the graph which shows the rate at which an enzyme works.

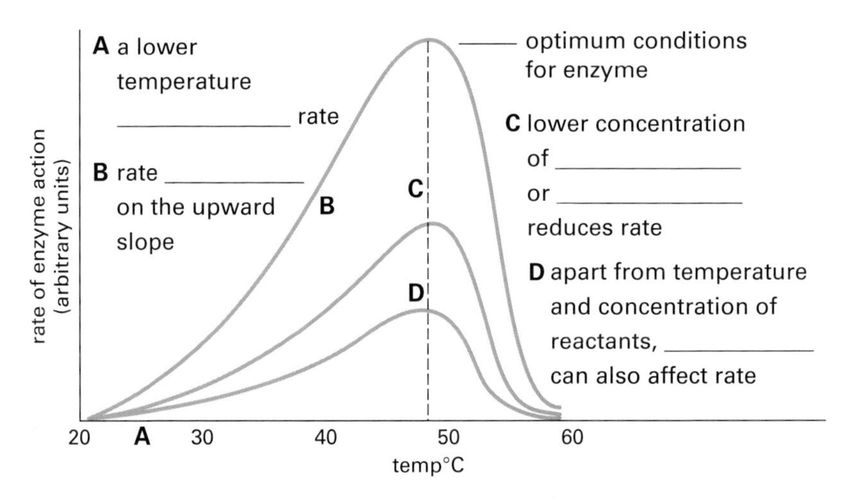

4 Complete the table describing each feature of the human circulation:

- direction of blood flow through the heart: ____________________
- direction of blood flow in arteries: ____________________
- reason for valves at the base of the large arteries: ____________________
- main tissue the heart is made of: ____________________
- importance of coronary arteries: ____________________
- what is meant by double circulation: ____________________
- the main components of blood: ____________________

5 Fill the blanks on the diagram of the heart to complete notes about its structure, and how it beats.

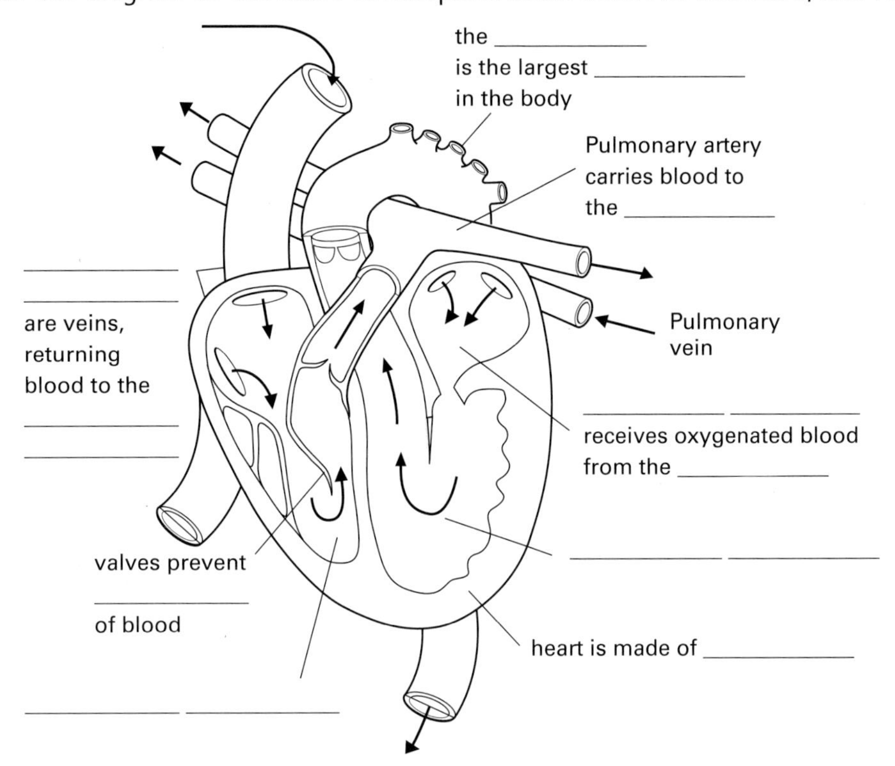

6 Complete the table on microbes.

name	example	disease it causes	medicines used to fight infection
virus			
bacterium			
fungus			

7 Complete the sentences about immune response:

An __________ is a particle or cell which is foreign to the body. These are detected by _______ ________ cells, which produce ______________ to match. The __________ lock onto the antigens, damaging them or allowing white blood cells to ___________ them. Some white blood cells make __________ which can neutralise toxins.

8 Complete the table of breathing events, choosing from the list in the last column.

	breathing in	breathing out	answers
ribs move			up/down
diaphragm			up /down
volume of chest			increases/decreases
air pressure inside chest			increases/decreases
air flows			out/in

9 a) Complete the equation for anaerobic respiration which may happen in the muscles during vigorous exercise: glucose → ___________ [+ ___________ transferred]

b) Complete the equation for anaerobic respiration which occurs in yeast cells:

glucose → ___________ + ___________ [+ ___________ transferred]

10 a) Complete the flowchart about the main parts of the nervous system using the following words:

spinal cord, sensory nerve, gland, brain, effector, receptor, motor, sensory, muscle, central

detecting

A ___________ neurone has a ___________ to detect a stimulus. Bundles of these neurones form a ___________, which carries impulses to the ___________ nervous system.

→

processing

The ___________ and ___________ ___________ are the main parts of the central nervous system. A reflex action may not involve the ___________.

→

responding

___________ neurones lead from the central nervous system to an ___________ e.g. a ___________ which produces a hormone or ___________, which causes movement.

b) Write a definition for each of the following terms:

synapse ___________

neurotransmitter ___________

connecting neurone ___________

11 Complete the table about the components of the eye and their functions.

part of the eye	function
cornea	
iris	
pupil	
ciliary muscle	
photoreceptor	
blood supply	
tear gland	

12 Complete the sentences:

Insulin is the hormone which ______________________________

Glycogen is a ______________________________

Glucagon causes ______________________________

Blood sugar level changes because ______________________________

13 Write a definition for each of the following terms:

ovulation ______________________________

menstruation ______________________________

implantation ______________________________

fertilisation ______________________________

14 Name one male and two female sex hormones.

15 Complete the table about the male and female reproductive systems.

part of the body	function
	produces sperms
	fertilised eggs develop here
scrotum	
	fuses with the egg cell during fertilisation
urethra	

16 Complete the flowchart describing events which happen when Tim works out at the gym:

Tim works out with weights, increasing ____________ in the muscles. This transfers more energy and body temperature ____________.

→

Tim needs to mop his forehead because ________ production increases. His face gets ________ because of increased ________ ________

→

Small ____________ near the skin surface widen, which is called ____________. Heat ____________ from the body, causing cooling near the skin surface.

17 Match each part of the kidneys with the part it plays in adjusting water content of the body.

part of the kidney	function
glomerulus	fine tunes salt and water balance, and helps regulate pH of blood
second convoluted tubule	concentrates urine by allowing reabsorption of water
blood supply around tubules	ultrafiltration of water and dissolved substances in the tubule
loop of tubule	allows reabsorption of glucose, amino acids and water back into blood
first convoluted tubule	puts reabsorbed materials back into circulation

18 Fill in the blanks to complete the labels.

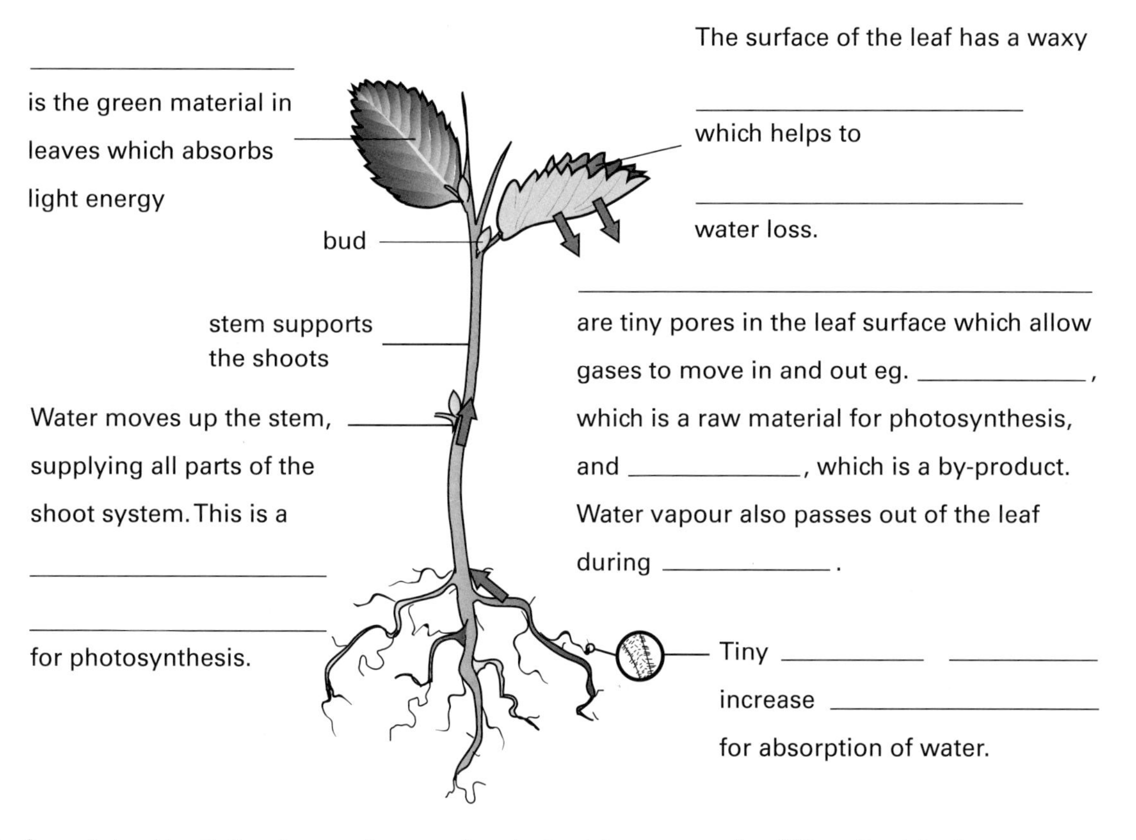

19 Complete the following sentences about plant hormones, by filling the blanks.

______________ are plant hormones which affect the growth of cells, by affecting the __________ of cell division and the amount of cell ___________. In ___________ the rate of cell growth increases where there is a higher concentration of hormone. This causes uneven growth on either side and a change of direction. Hence a shoot may grow towards light. This type of growth movement is called a __________.

20 Complete the table about types of reproduction by writing true or false in a box.

	asexual	sexual
two parent cells are involved		
mitosis is the type of cell division which occurs		
cloning is an example of this type of reproduction		
there is genetic variation		
sex cells are produced		

21 The people named below developed theories of evolution. Match statements which are true for each person's theory by drawing a line from the statement to the name.

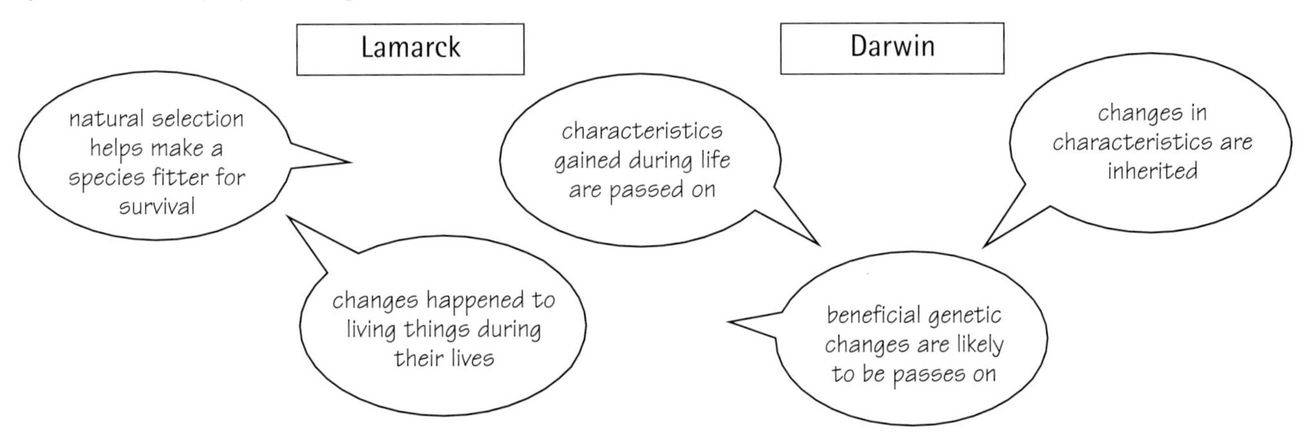

22 Complete the Punnet square, showing inheritance of red flower colour (R) and white (r) flower colour.

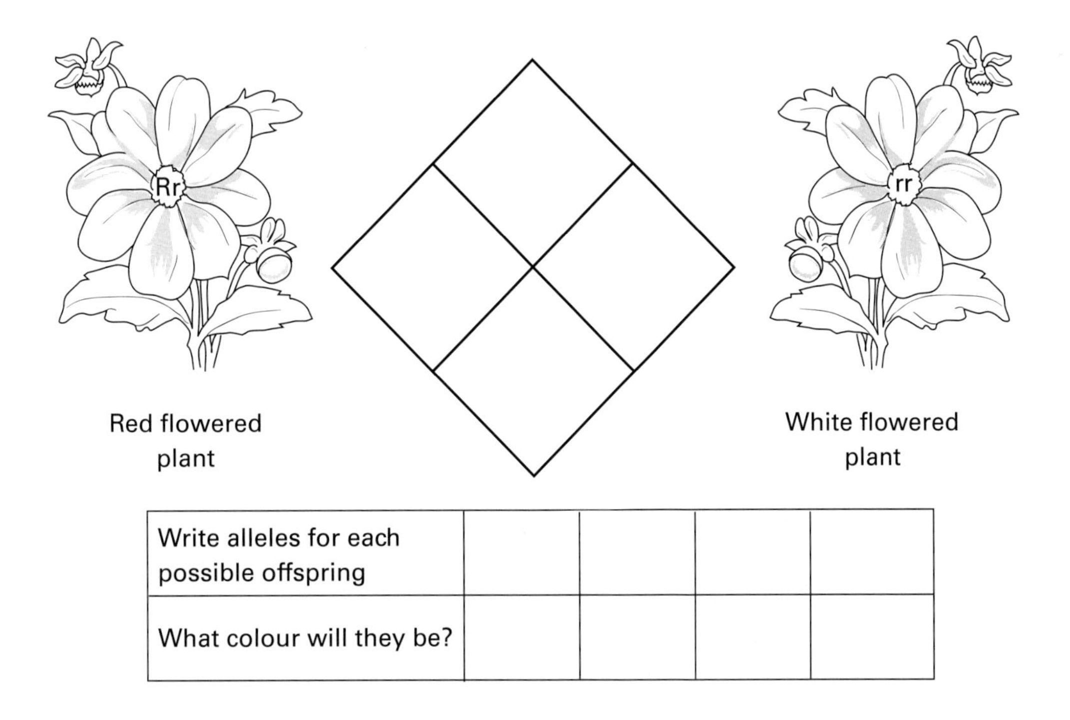

Write alleles for each possible offspring				
What colour will they be?				

23 Name five types of pollution which may be caused by human activities.

________________ ________________ ________________

________________ ________________

24 Write a food chain which might occur in:

a woodland ________ → ________ → ________ → ________

the desert ________ → ________ → ________ → ________

the sea ________ → ________ → ________ → ________.

25 Are these statements true or false? Write true or false next to each one.

Carbon dioxide enters the atmosphere as a result of: a) respiration ________

b) transpiration ________

Oxygen is a by-product of: a) fermentation ________

b) photosynthesis ________

Burning fossil fuels produces: a) carbon dioxide ________

b) nitrogen oxides ________

Chemistry

1 Label the diagram of an atom.

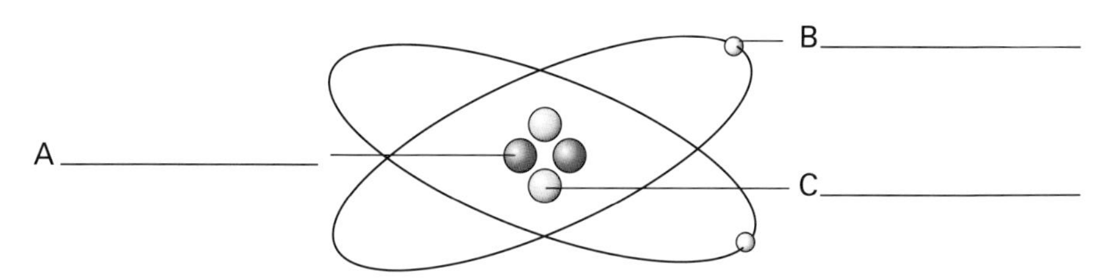

2 Complete the following table of numbered elements.

Element	Mass number	Atomic number	No of protons	No of neutrons	Number of electrons	Electronic configuration
1	27	13				
2			15	16		
3		17		19		
4	40				18	

3 Draw the shells and electrons and put the charges in the boxes on this diagram of the formation of sodium chloride.

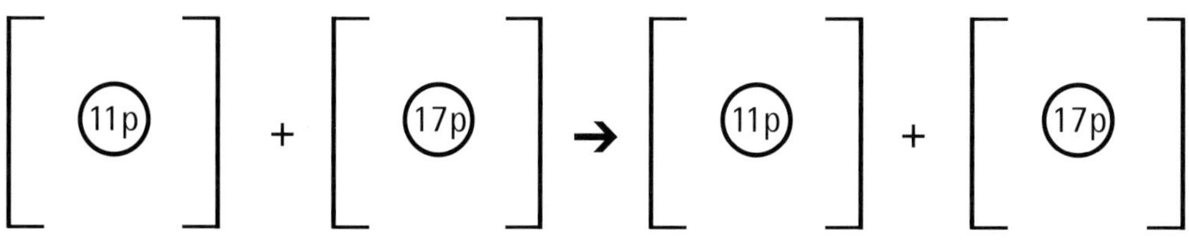

a Na atom a Cl atom a Na ion a Cl ion

4 Draw the outer orbits and the electrons on the following covalent molecules

a) water b) ammonia c) methane

```
   O                                  H
H     H        H  N  H             H  C  H
                  H                   H
```

5 Complete and balance the following equations:

a) $Mg(s) + O_2(g) \rightarrow$ ____________________

b) $Zn(s) + HCl(aq) \rightarrow$ ____________________

c) $CaCO_3(s) + HCl(aq) \rightarrow$ ____________________

d) $Na(s) + H_2O(l) \rightarrow$ ____________________

e) $Cl_2(g) + NaBr(aq) \rightarrow$ ____________________

6 Fill in the working and the answers in the following calculation of the percentages of the elements present in sodium sulphate (Na_2SO_4), if O = 16, Na = 23, S = 32.

- relative molecular mass = = ________
- percentage of sodium = = ________
- percentage of sulphur = = ________
- percentage of oxygen = = ________

7 Complete the following:

i) acid + base $\rightarrow$ ____________________

ii) acid + metal $\rightarrow$ ____________________

iii) acid + carbonate $\rightarrow$ ____________________

8 Write 'strong acid', 'strong alkali', 'weak acid', 'weak alkali', 'neutral', in the appropriate place on the pH scale.

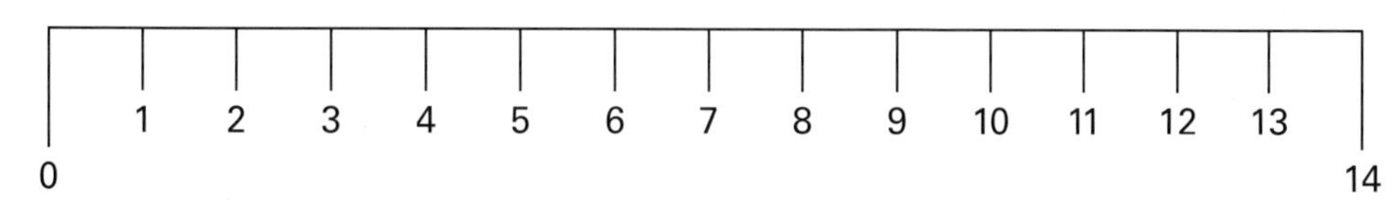

9 Put oxidation/reduction arrows on each of the following redox reactions. Also write on the arrow what is lost or gained.

a) $C\ (s) + ZnO\ (s) \rightarrow Zn\ (s) + CO\ (g)$

b) $H_2S\ (g) + Cl_2\ (g) \rightarrow 2HCl\ (g) + S\ (s)$

c) $Mg\ (s) + Fe^{2+}\ (aq) \rightarrow Mg^{2+}\ (aq) + Fe\ (s)$

10 Complete the following table.

name	ethane	ethene	chloroethane
molecular formula			
structural formula			

11 Complete the details for the electrolysis of aluminium oxide to produce aluminium.

ions present: Al_2O_3 →

at anode (+): [ ]

at cathode (–): [ ]

overall: Al_2O_3 → [ ]

12 Complete the table of the present composition of the atmosphere.

substance	%
nitrogen	
	21
noble gases	
	0.03
H_2O vapour	

13 Write in the four missing items on this diagram of the rock cycle.

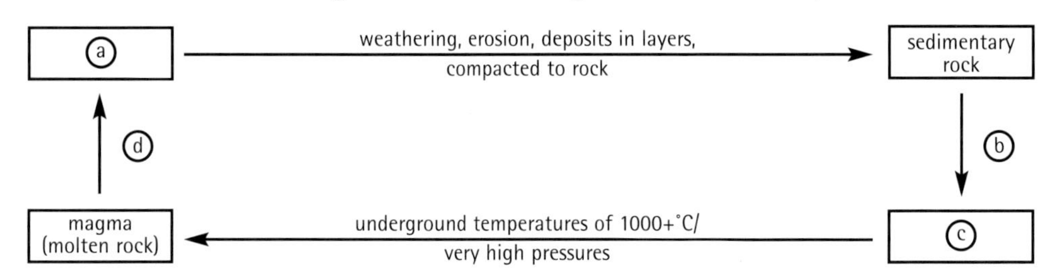

14

I	II	III	IV	V	VI	VII	0
A						D	
		B			C		E

(The letters are NOT the symbols of the elements)

The diagram shows periods 2 and 3 of the periodic table. In the table below give at least two facts about each of the elements A to E.

element	facts
A	
B	
C	
D	
E	

15 Fully label the energy level diagram below.

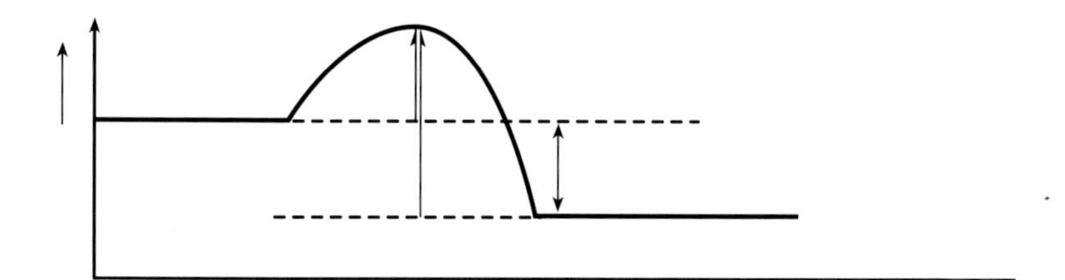

16 Complete the equation for the Haber process for the manufacture of ammonia.

$$N_2\,(\ \ ) + \quad (g) \underset{\text{catalyst}}{\overset{\text{atm/}\qquad {}^{\circ}\text{C}}{\rightleftharpoons}} 2 \quad (g);$$

ΔH is

Physics

1 Use the variables in the centre to make equations

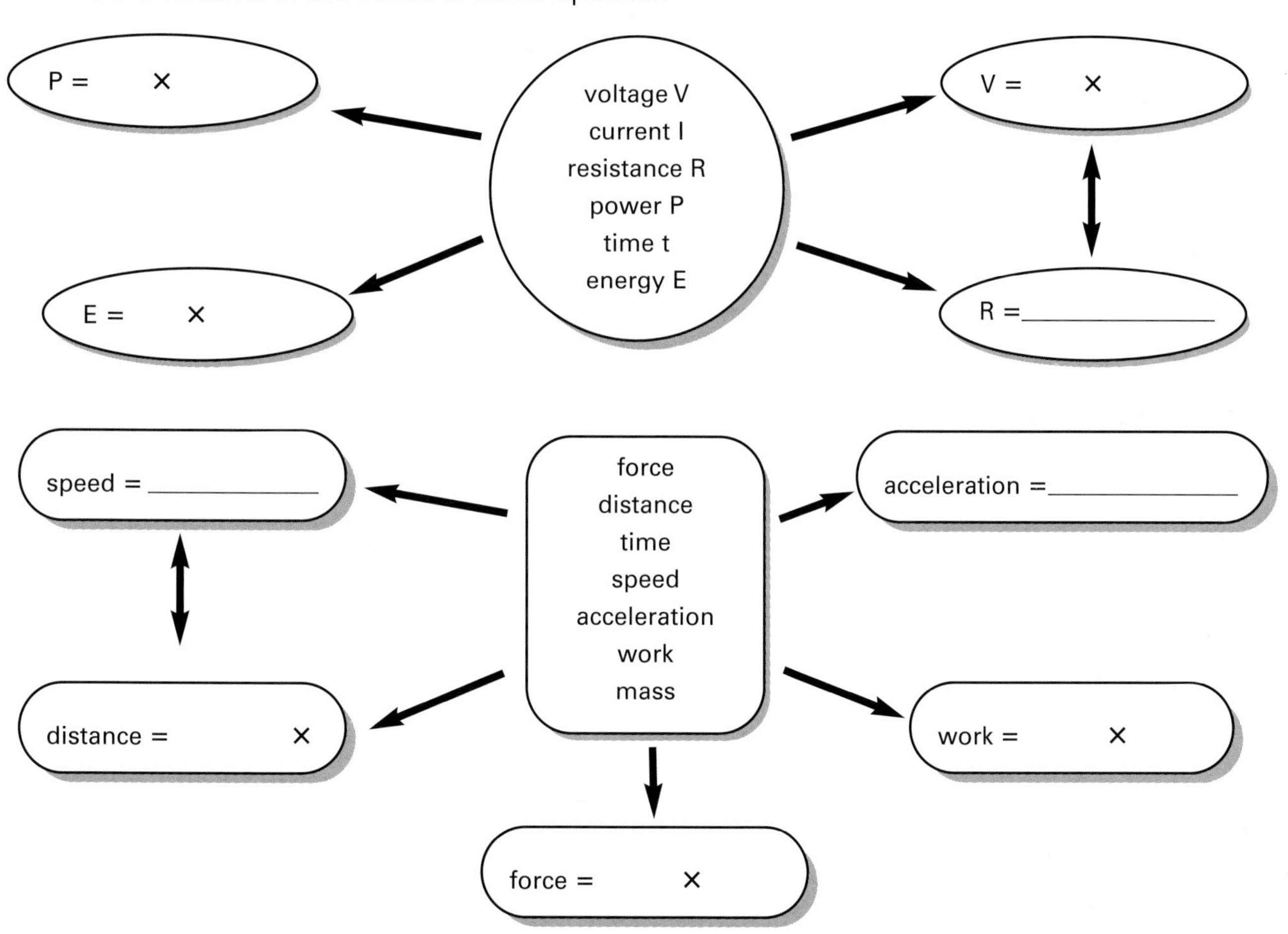

2 Complete the sentences

a) The resistance of metals increases as the temperature ______________. The resistance of a thermistor ______________ as its temperature increases. The resistance of an LDR ______________ as the light level increases.

b) The primary coil of a transformer is connected to an ____________ current supply. As the current in the primary coil varies it sets up a _______ ________ field in the iron core, which ________ a changing voltage in the secondary coil.

3 Fill in the gaps in the tables below on waves:

electromagnetic spectrum	gamma rays		ultraviolet		infrared		radio waves
use		medical imaging		seeing		cooking and communication	

wave	**oscillation direction**	**examples**
longitudinal		
transverse		

4 Name the planets in order from the Sun:

Mercury				asteroid belt					Pluto

5 Fill in the gravity gaps below:

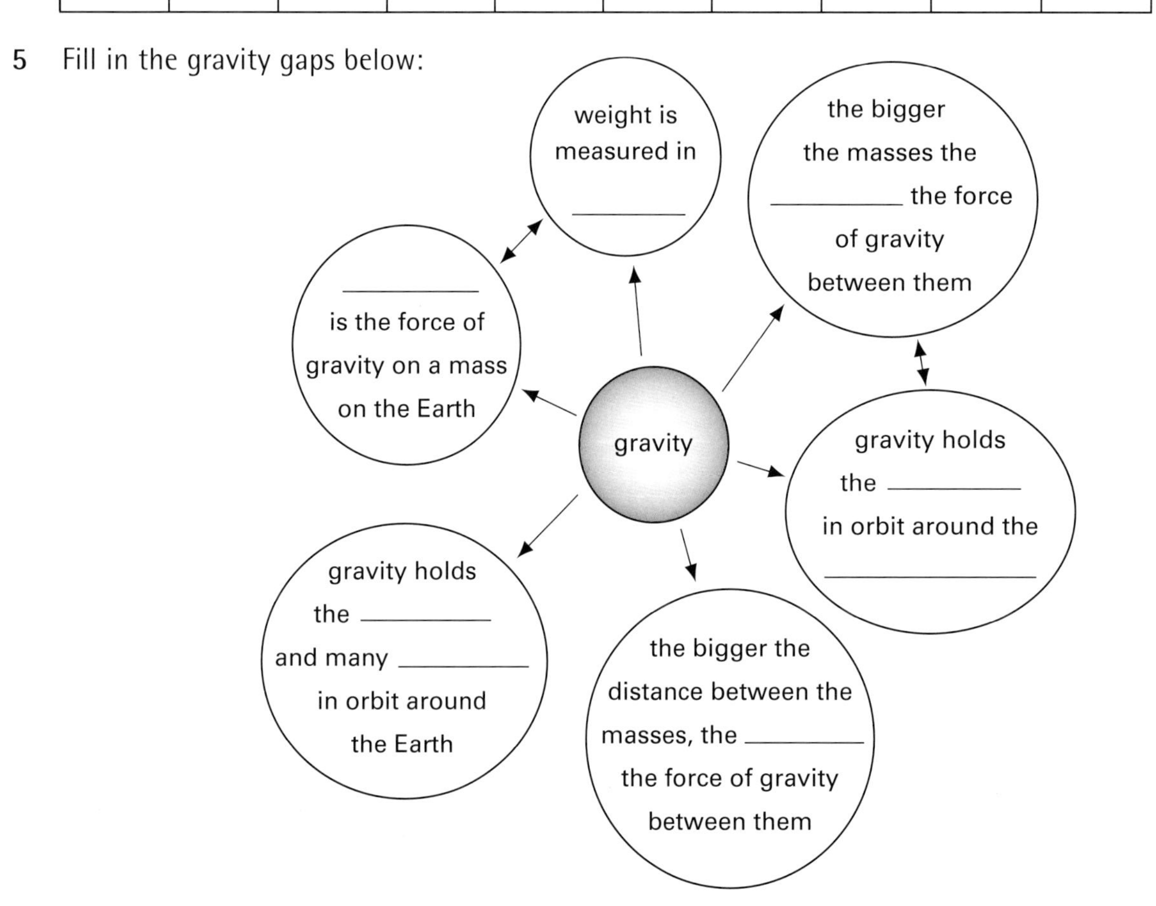

6 Complete the life story of a star:

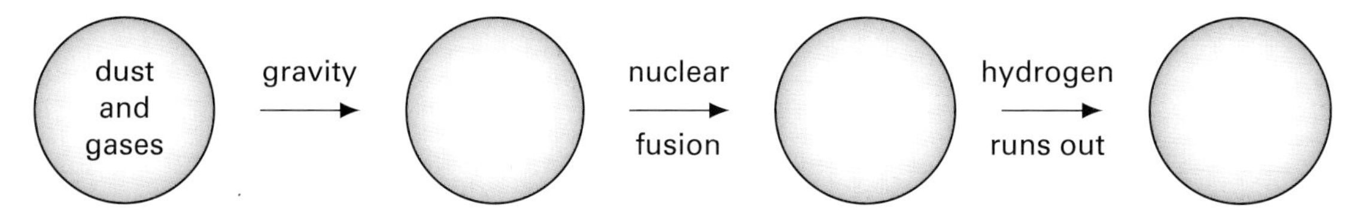

7 Complete the table on energy transfer:

method of transfer	what happens in the material	example
	energy of vibrating particles is transferred to neighbouring particles	metals are good conductors and are used to make saucepans
convection		
radiation		energy reaches from the sun this way
		sweating removes energy and prevents the body from overheating

8 Fill in the gaps about radioactivity:

radiation	what is it	charge?	how far can it travel in air?	what stops it?
alpha α				
beta β				
gamma γ				

Biology answers

1 1 movement in/out of cell 2 cytoplasm 3 nucleus 4 tissue 5 organ, digestive

2 egg = protein butter = fat/lipid bread = carbohydrate (starch)
in mouth: starch, carbohydrase/amylase, sugar/glucose
in stomach: protein, protease (acid helps), amino acids
in small intestines: all, absorbed

3 A lower temp **reduces** rate; B rate **increases** on upward slope; C **substrate** or **enzyme**; D pH

4
- into atria → ventricles → arteries
- from heart to organs
- prevent back flow to atria
- muscle
- supply oxygen/food to heart muscle
- RHS to lungs; LHS to most of body
- cells, plasma, platelets, dissolved materials

5 clockwise from top right: aorta; artery; lungs; right atrium/ auricle; lungs; left ventricle; muscle; right ventricle; backflow; venacava; left atrium

6

name	example	disease	medicines
virus	HIV	AIDS	anti-virals
bacterium	streptococcus	sore throat	antibiotic
fungus	various	athletes foot	antifungal

7 antigen, white blood, antibodies, antibodies, engulf, antitoxins

8

breathing in	breathing out
up	down
down	up
increases	decreases
decreases	increases
out	in

9 a) glucose → lactate [+ energy transferred]
b) glucose → ethanol + carbon dioxide [+ energy transferred]

10 a) **detecting:** sensory, receptor, sensory nerve, central
processing: brain, spinal cord, brain
responding: motor, effector, gland, muscle
b) **synapse** – junction between two nerve cells; **neurotransmitter** – chemical which diffuses across synapse/triggers impulse in adjacent neurone; **connecting neurone** – connects sensory and motor neurones/found in central nervous system

11 **cornea**: helps focus light/form image
iris: controls amount of light entering eye
pupil: allows light to enter eye
ciliary muscle: adjusts tension on ligaments holding lens/thickness of lens
photoreceptor: detects light

blood supply: supplies eye with oxygen
tear gland: washes eyes with tears/prevents infection/antibacterial

12 insulin: converts glucose to glycogen; glycogen: is a carbohydrate store; glucagon causes conversion of glycogen to glucose; intake/what we eat varies

13 ovulation: release of eggs from ovary; menstruation: monthly loss of lining of womb/blood; implantation: fertilised embryo becomes attached/embeds in lining of womb; fertilisation: nucleus of two sex cells/sperm and egg fuse

14 testosterone; oestrogen, follicle stimulating hormone/FSH, progesterone

15 **testis**: produces sperm
uterus: fertilised eggs develop
scrotum: **supports testes**
nucleus of sperm: fuses with the egg during fertilisation
urethra: **passes urine from bladder out of body**

16

respiration	sweat	capillaries
increases	flushed/red	vasodilation
	blood	radiates
	circulation	

17 **glomerulus** ultrafiltration of water and dissolved substances in the tubule **second convoluted tubule** fine tunes salt and water balance, and helps regulate pH of blood
blood supply around tubules allows reabsorption of glucose, amino acids and water back into blood
loop of tubule concentrates urine by allowing reabsorption of water
first convoluted tubule puts reabsorbed materials back into circulation

18 clockwise from top right: cuticle, reduce; stomata, carbon dioxide, oxygen, transpiration; root hairs, surface; raw material; chlorophyll.

19 auxins, rate, elongation/growth, stems/shoot tips, tropism

20 sexual; asexual; asexual; sexual; sexual

21 Natural selection helps make a species fitter for survival. (Darwin); Characteristics gained during life are passed on. (Lamarck); Changes in characteristics are inherited. (Darwin); Beneficial genetic changes are likely to be passed on. (Darwin); Changes happen to living things during their lives. (Lamarck)

22

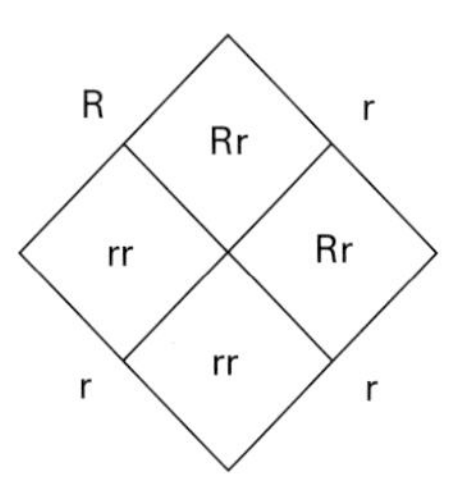

Write alleles for each possible offspring	Rr	Rr	rr	rr
What colour will they be?	Red	Red	White	White

23 oxides of nitrogen/sulphur/acid rain; carbon dioxide/global warming; pesticides/agro-chemicals in waterways; nuclear waste; dumping rubbish; methane from landfill sites; light at night etc

24 **woodland:** tree → squirrel → fox → decomposers;
desert: cactus → insect → lizard → vulture;
sea: plankton → small fish → large fish → humans

25 true, false, false, true, true, true

Chemistry answers

1 A electron
B proton (or neutron)
C neutron (or proton)

2 missing numbers are: 1) 13 14 13 2.8.3 2) 31 15 15 2.8.5
3) 36 17 17 2.8.7 4) 18 18 22 2.8.8

3

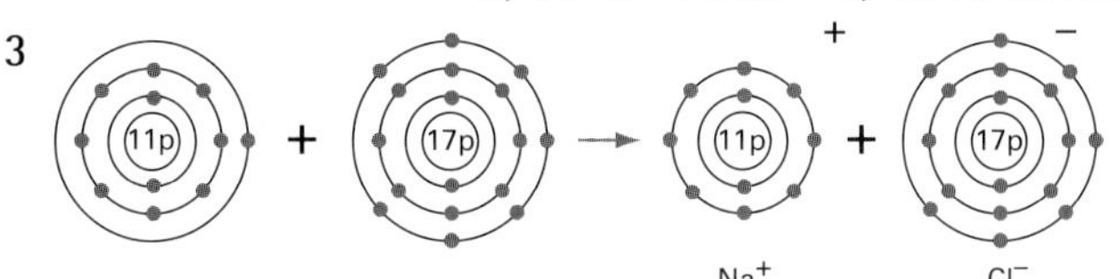

4

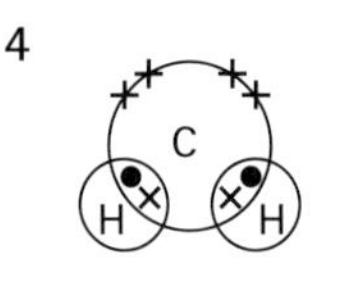

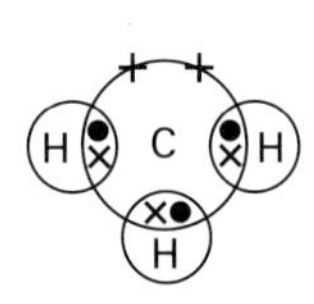

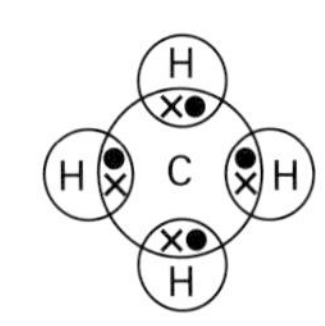

5 a) $2Mg\ (s) + O_2\ (g) \longrightarrow 2MgO\ (s)$
b) $Zn\ (s) + 2HCl\ (aq) \longrightarrow ZnCl_2\ (aq) + H_2\ (g)$
c) $CaCO_3\ (s) + 2HCl\ (aq) \longrightarrow CaCl_2\ (aq) + H_2O\ (l) + CO_2\ (g)$
d) $2NA\ (s) + 2H_2O\ (l) \longrightarrow 2NaOH\ (aq) + H_2\ (g)$
e) $Cl_2\ (g) + 2NaBr\ (aq) \longrightarrow 2NaCl\ (aq) + Br_2\ (l)$

6 relative molecular mass $= (23\times2) + 32 + (16\times4) = 142$

% Na $= \frac{46}{142} \times 100 = 32.4$

% S $= \frac{32}{142} \times 100 = 22.5$

% O $= \frac{64}{142} \times 100 = 45.1$

(OR: 100 – 32.4 – 22.5)

7 i) ⟶ salt + water ii) ⟶ salt + hydrogen
iii) ⟶ salt + water + carbon dioxide

8

0	1	2	3	4	5	6	7	8	9	10	11	12	13	14
	strong acid				weak acid		neutral	weak alkali				strong alkali		

9 a)

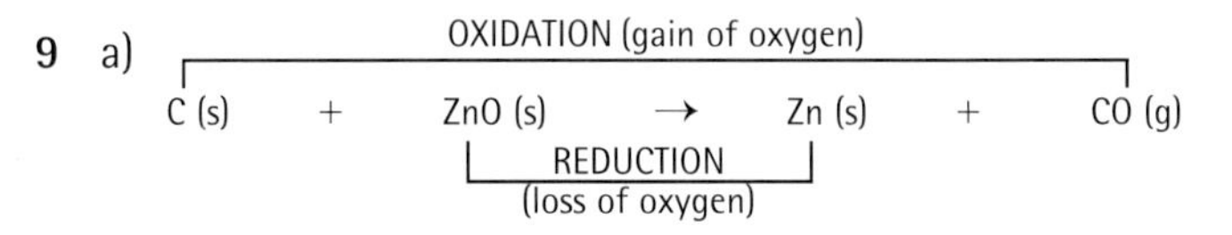

b)

OXIDATION (loss of hydrogen)
$H_2S\ (g) + Cl_2\ (g) \rightarrow 2HCl\ (g) + S\ (s)$
REDUCTION
(gain of hydrogen)

c) OXIDATION (loss of electrons)

$Mg\ (s) + Fe^{2+}\ (aq) \rightarrow Mg^{2+}\ (aq) + Fe\ (s)$

REDUCTION (gain of electrons)

10 ethene ethane chloroethane

C_2H_6 C_2H_4 C_2H_5Cl

```
  H H        H     H       H H
  | |         \   /        | |
H-C-C-H        C=C       H-C-C-Cl
  | |         /   \        | |
  H H        H     H       H H
```

11 ions present: $Al_2O_3 \rightarrow 2Al^{3+} + 3O^{2-}$

at anode (+): O^{2-} ions are attracted

$3O^{2-} - 6e^- \rightarrow 3O$: oxidation

$3O + 3O \rightarrow 3O_2\ (g)$

at cathode (–): Al^{3+} ions are attracted

$2Al^{3+} + 6e^- \rightarrow 2Al\ (s)$: reduction

overall: $2Al_2O_3 \xrightarrow[\text{decomposition}]{\text{electrolytic}} 4Al\ (s) + 3O_2\ (g)$

12 nitrogen: **78%** **oxygen**: 21 noble gases: **1%**

carbon dioxide (CO_2): 0.03% H_2O vapour: **variable**

13 a) igneous rock b) heat and high pressure, deep underground c) metamorphic rock d) magma rises, cools quickly or slowly

14 There are a number of possible answers for each, of which students are required to choose at least two. The answers must be based on the periodic table group of the element, i.e. A: group I (alkali metal) B: group III C: group VI D: group VII (halogen) E: group 0 (noble gas)

15

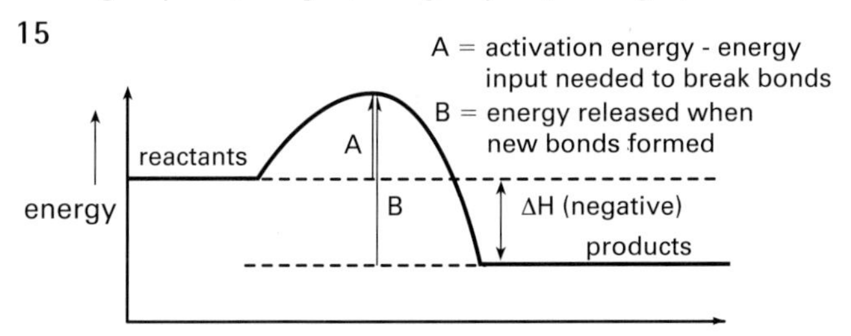

16

$$N_2\ (g) + 3H_2\ (g) \underset{\text{Fe catalyst}}{\overset{\text{250 atm/500-550°C}}{\rightleftharpoons}} 2NH_3\ (g);$$

ΔH is negative

Physics answers

1 $P = V \times I$; $V = I \times R$; $E = P \times t$; $R = V / I$

speed = distance / time;

acceleration = change in speed / time taken;

work = force × distance;

distance = speed × time;

force = mass × acceleration

2 a) increases; decreases; decreases

b) alternating; changing magnetic; induces

3 reading across: x-rays, visible light, microwave, treatment of cancer, sterilising medical equipment, heating, communications

longitudinal	in same direction as energy travels	sound waves, P earthquake waves
transverse	at right angles to direction energy travels	light, water waves, earthquake S waves

4 Mercury, Venus, Earth, Mars, asteroid belt, Jupiter, Saturn, Uranus, Neptune, Pluto

5 **Weight** is the force of gravity on a mass on the Earth

Weight is measured in **newtons**

The bigger the masses the **bigger** the force of gravity between them

Gravity holds the **planets** in obit around the **Sun**.

The bigger the distance between the masses, the **smaller** the force of gravity between them

Gravity holds the **Moon** and many **satellites** in obit around the Earth.

6 very hot gases, adult star, red giant

7 conduction

warmer, less dense liquid or gas rises above denser, older fluid; energy from the element at the bottom of a kettle is transferred to all the water.

electromagnetic waves (infrared); energy reaches us from the Sun this way.

evaporation; particles evaporating from a liquid carry away energy, leaving the liquid cooler

8

alpha α	2 protons 2 neutrons	positive	few centimetres	paper
beta β	electron	negative	few metres	aluminium
gamma γ	electro-magnetic wave	none	a long way	thick lead, very thick concrete

For each topic: answers are given to the questions on the borders of the pages, in the order they appear on the spread, followed by answers to the practice questions.

Life processes

A being aware of/responding to danger, selecting better environments, finding food/a mate
A many different cell types adapted for different functions/life processes more varied
A muscle
1 permanent increase in size, due to increase in size/number of cells
2 feeding/nutrition
3 to transfer energy form food to ATP/cells/body; to get energy for life processes
4 carbon dioxide in exhaled air; urea in urine/sweat
5 light sensitive cell, nerve tissue, eye

Cells and cell activities

A nucleus acts as a set of instructions for cell processes, including inheritance of characteristics; chloroplast is where photosynthesis happens in plant cells; respiration happens in mitochondria
A job is to carry out photosynthesis, related structure is that it contains many chloroplasts which absorb light energy
1 nucleus
2 cellulose cell wall and vacuole containing watery sap
3 a) both involved in absorption
b) diffusion; osmosis (diffusion of water); active transport
4 chloroplast containing chlorophyll, which uses light energy to build glucose molecules

Body systems

A to transport a wide variety of materials within the organism because diffusion alone is not fast enough
A leaf, contains layer of palisade cells, these have many chloroplasts containing chlorophyll
1 leaves get more light if they grow above other plants; flowers more conspicuous for insect pollination and higher into air currents for wind pollination
2 it absorbs nutrients and water though the roots and these are transported around the plant by xylem tissue
3 in the flower: anther in male and ovary in female
4 blood system: white blood cells produce antibodies specific to antigens
5 nervous system - sends impulses; skeleto-muscular - moves in response; blood - transports food and oxygen to muscles; breathing system - gets oxygen into the body.

Human nutrition

A sugar passes directly into bloodstream and excess is converted to glycogen and stored; starch must be digested, which happens over a period of time so there is a slower absorption of glucose into the bloodstream.
A spinach; meat
A generally, men have larger body mass/more muscle tissue
1 carbohydrate (sugar/starch); fat/lipid
2 likely to contain more vitamins than highly cooked foods; contain fibre
3 protein
4 eating too much, eating too much high calorie food e.g. fat and sugar, eating unbalanced diet, lack of exercise

Human digestion

A liver
A peristalsis
A pH in stomach is too low/too acid
1 salivary gland; pancreas; lining of the small intestine
2 long length; inner surface has finger-shaped villi; cell surfaces have microvilli
3 shape of protein changes/active site destroyed
4 a) substrate must bond with active site on enzyme molecule for a reaction to happen, hence less enzyme molecules mean less active sites and the rate of reaction is lower
b) pH is not optimum

The heart and heart beat

A left atrium, right atrium, left ventricle, right ventricle, aorta, vena cava
A heart muscle needs large supply of O_2 as it is very active tissue; blockage stops circulation to heart muscle itself
1 LHS and RHS work independently but simultaneously; RHS pumps to lungs and back to heart, LHS pumps to rest of body and back to heart
2 contracts in a wave/atria contract then ventricles/squeeze blood out into main arteries
3 valves are between atria and ventricles – close when ventricles are full and begin to contract; valves at base of large arteries – close when ventricles stop contracting

Blood and circulation

A one circulation goes to the lungs and returns to the heart; the other pumps around the rest of the body
A O_2 is constantly used by cells during respiration, hence concentration reduces unless there is further supply; and CO_2 is produced during respiration, hence concentration increases unless it is removed
1 aorta receives blood at higher pressure as left ventricle contracts, while vena cava collects blood which is some distance from the heart; aorta contains oxygenated blood from the left side of the heart, but blood in vena cava has passed through tissues and oxygen has diffused into cells
2 arteries more muscular and alter in size to alter blood pressure; elastic to allow return to shape and cope with higher blood pressure; veins have larger lumen so blood pressure is lower, but valves prevent back flow of blood; capillaries have very thin walls to allow easy exchange of materials
3 diffusion gradient/difference in concentration of oxygen between red blood cells and tissue cells
4 produce antibodies; engulf non-self/foreign proteins

Health and disease

A earache = infection; broken leg = injury; additions to tranquilisers = lifestyle habit; feeling seriously depressed = mental illness; haemophilia = inherited
A virus, bacterium, fungus
1 an infection
2 diet, exercise, avoid drug abuse (including smoking)
3 if the incidence of lung cancer is greater for smokers than non-smokers
4 the medicine has been safety tested, doses and effects known; drug purity ensured

Body defences and immunity

A skin may be damaged/injured allowing entry of microbes
A injury can result in bleeding to death
A white blood cells
A causes an immune response; immune memory
1 immunisation
2 produce antibodies; engulf/destroy germs/microbes/bacteria; neutralise toxins

The breathing system

A trachea ➔ bronchi ➔ bronchioles
A oxygen is continually used in tissues for respiration, so has to be replaced active process/muscles contract to lift ribs/flatten diaphragm
A tars; nicotine; carbon monoxide
1 many of them, large surface area, thin-walled for fast diffusion, rich blood supply
2 by changing volume of chest cavity, which changes air pressure inside lungs compared to outside
3 air sac damaged, mucus collects in lungs, less oxygen absorbed from air

Transferring energy by respiration

A to transfer energy to cells
A muscles need energy, use muscles more when exercising, so need greater air intake
A gaseous exchange and supply of O_2 to cells does not keep up with demand
A ratio of body surface area to body mass is greater/more surface to lose heat
1 aerobic uses oxygen, anaerobic doesn't; aerobic transfers more energy than anaerobic; glucose completely broken down to CO_2 and water during aerobic, only partly broken down in anaerobic.
2 When there is a build up of lactate/demand for oxygen is greater than supply
3 1 $cm^3/g/h$

Sensitivity - detecting change

A photoreceptors in retina, touch receptors in skin; sound receptor in ear
A photoreceptors in retina, touch receptors in skin;
A impulses cannot pass along sensory nerves to brain, so unaware of pain
A interprets information and decides on action
1 one end is developed as a receptor/can detect stimuli/changes
2 insulate the axon

Sensitivity - response to change

A a) no need to think about processes such as heart beat and breathing which are vital for life
b) allow organisms to avoid/get out of danger fast
A alcohol alters judgement; is a sedative: slows responses and causes drowsiness
1 reflex: moving bare foot off sharp stone; crying when peeling onions; catching onto a rail to save yourself from falling (others are voluntary)
2 could paralyse muscles involving heart beat or breathing, which must move continuously

The eye and sight

A eyelid closes over eye surface, blinking reflex, eye sunk deep into bony socket, eyelashes catch dirt
A to allow change in shape necessary for focusing
A photoreceptors in retina ➔ sensory neurones in optic nerve to brain ➔ motor neurones to iris
1 antibacterial tears, eyelashes, eye lid
2 middle layer/choroid
3 takes information from receptor cells to the brain
4 brain interprets the image

Chemical control in the body

A situated at base of brain/controls other glands in the bloodstream
1 via the blood stream
2 ovary; uterus/womb
3 beard growth, development of testes/sperm production; body mass increases/muscle development; pubic hair

Controlling sugar level in blood

A glucose = a ready food supply/needed for respiration; glucagon = hormone to convert glycogen to glucose; glycogen = food store/can be used when blood sugar levels drop; insulin = converts excess glucose to glycogen/helps prevent cell damage
A genetically-engineered insulin is same as human insulin so no rejection problems; can produce large amounts cheaply and quickly; can control purity and concentration more easily
A level of glucose/sugar in blood
1 blood sugar would increase rapidly, and so would rate of respiration; insulin level rises to cope with sudden increase
2 a) keep sugar intake fairly constant
b) to avoid rapid increase/decrease in blood sugar level

Sexual maturity

A testosterone; progesterone, oestrogen and FSH (follicle stimulating hormone)
A 14
A In case the sperms get passed the cap or condom
1 39 years x 12 eggs per year = 468
2 lack of FSH
3 avoiding sexual intercourse around the time of ovulation/'rhythm' method

Body systems for reproducing

A a) testis b) epididymis
A placenta develops here, which supplies foetus with nutrition/oxygen
A not all eggs can be fertilised; eggs may die after fertilisation; implantation unsuccessful
1 a) ovary: produces eggs b) oviduct: eggs may be fertilised as they pass along here c) uterus: a secure place for the foetus to develop d) vagina: sperms may enter through here
2 rate of fertilisation is higher/eggs more likely to survive
3 a) thought it was her fault/her body wasn't working as it should/didn't feel like a 'real' woman without children/ wanted to provide children for partner
b) unable to get on with the rest of life; cause tension between them if one partner wants children more than the other; place a financial strain; one partner blames another if treatment is unsuccessful
c) man might not ever completely accept child as his own/feel a failure for not producing enough/viable sperm

Keeping the body in balance

A skin damage allows fluid loss, and infection
A increase, due to greater fluid intake

A able to regulate body temperature independently of environment
1 a) blood sugar/glucose level b) glands: pancreas produces insulin; liver converts excess glucose to glycogen
2 drink plenty of water to replace that lost in sweat
3 blood vessels constrict/dilate; sweat glands produce sweat
4 muscle

Plants and photosynthesis

A smaller raw material molecules are built into larger carbohydrate molecules
A between 0.03 and 0.1% lines
1 a) CO_2 diffuses through stomata; water absorbed by roots and is transported in xylem b) in phloem
2 actively growing/for cell division/building new tissues
3 upward slope
4 supplying more CO_2 increased the rate above the 0.03% line/1.0% line is higher
5 lack of chlorophyll to transfer solar energy

Water and transport implants

A solute is a substance which is dissolved in a liquid. A solvent is a liquid which dissolves a solute. A solution is a mixture: a liquid which has a solute dissolved in it.
A wet, still/calm air, cold (also dark)
A organic – waste from or remains of living things e.g. manure, compost; inorganic – manufactured chemically in factories
1 by evaporation through the stomata
2 by osmosis
3 diffusion
4 most of the stomata are on underside of leaf, so grease blocked them in leaf A and less water was lost

Controlling plant growth

A growth movements in response to stimuli
A stems grow more rapidly until leaves reach light; help stems grow towards light by stimulating growth on shady side
1 shoots grow away from gravity, roots grow towards gravity
2 auxin is a plant hormone; chemical analysis
3 increases growth

Plant hormones in action

A not all evidence/data is available at one time/research continues to produce new evidence
A want to kill weeds e.g. dandelions, not grasses
1 concentration of auxin same in lit and shaded shoot tips, but distribution is different
2 growth slows on the lower side of the root and root curves downwards
3 to stimulate growth of side shoots/make a bushier shape

Variation and genetics

A each puppy inherits a slightly different combination of genes from each sex cell
A a) 23 b) 46 (or 2 × 23 or 23 pairs)
A advantage: know what problems might arise/predict treatments; disadvantage: might influence abortion rate/affect people's will to live
1 during cell division when sex cells are formed; due to mutation
2 some characteristics are beneficial and give an organism an advantage; others are harmful or even lethal

Evolution and genetic engineering

A **creation**: God created Earth; **acquired characteristics**: happened during lifetime of organism, but could be passed on; **natural selection**: new characteristics arise by change to DNA, advantageous characteristics more likely to be inherited as those individuals are more likely to survive and reproduce
A safe, cheap, plentiful
A growth hormone
1 a) 6% in unpolluted area, 53% in polluted area b) how easily they are spotted by predators/how well camouflaged they are c) decrease in dark and increase in lighter moths
2 supply a lacking gene in nasal spray/produce medicine

How genes are passed on

A all three buds have the same genes; no genetic variation; no sex cells involved
A a) sex cell produced by meiosis which halves the number of chromosomes b) baby's cells have two sets of chromosomes, one from each parent
A to breed animals with desirable combinations of characteristics
1 sexual reproduction: b and c asexual reproduction: a
2 a) high milk yield; creamy milk b) rapid growth rate

Patterns of inheritance

A by observing characteristics and using genetic diagrams
A red and white flower colour remained separate/no blending/no pink flowers
A **homozygous** individuals have to identical alleles for a particular characteristic, **heterozygous** individuals have a recessive and a dominant allele for a particular characteristic
1 red colour appeared to hide white colour in first generation of offspring
2 3 red :1 white
3 50%

More about inheritance

A balance between genders is optimum for human population maintenance
A 20%
A 0% (or no) chance
1 a) 2 recessive alleles b) because the females produce sons with haemophilia c) 20% d) because the disease results from a deficiency in the coltting factor
2 a) Rumina has freckles, which is a recessive trait b) none of them have freckles c) Ff, heterozygous for the condition but phenotype is without freckles d) neither parent has freckles but an offspirng does, so each must have provided an f allele

Ecosystems and biodiversity

A aquatic/marine
A is a vast gene pool; provides materials/resources/useful substances
1 environment, organisms, interactions between them
2 to maintain the gene pool/cannot predict the longer term effects of reducing biodiversity
3 a) space, light, nutrients b) less light near hedge/hedge uses more of the available nutrients d) use fertiliser to improve growth/trim or remove hedge

Human impact on environment

A increasing human population
A a) fisherman has to earn a living, and wants to catch fish while protecting stocks longer term; b) shopper is interested in choice of fish and price c) scientists want to help prevent loss of fish populations/help find ways of increasing fish stocks d) politicians have to explain management policies

to the public e) farmers may farm fish/may also compete with fishermen for food sales

1 human activities have an impact due to use of resources and waste production
2 a resource which cannot be replaced
3 competing priorities means that each stakeholder group have a different interest e.g. scientists investigate factors affecting fish populations and 'knock-on' effects/how to protect species; fishermen need to earn a living now and in the future
4 produces CO_2, which causes global warming

Survival and interdependence

A needs to conserve water
A producers = lillies, bull rushes, plants; food chains e.g. plants ➔ insect ➔ frog; plants ➔ ducks; plants ➔ little fish ➔ big fish etc.
1 having characteristics which enhance survival in a particular environment
2 heat loss is greater for an small animal with larger surface area to body mass; need to eat more/feed more continually
3 less water loss
4 population of fish which eat beetles might reduce; fish might eat more other organisms and their populations might reduce

Predators, prey and resources

A uses natural predator of a pest to control the pest population
A nutrients are recycled/dead remains are returned to ecosystem
1 predator population mimics prey, but after a time lag/predator depends on prey for food, size of prey population depends on number of predators
2 no natural predators for the cane toads so population increased dramatically, killing many other animals apart from rats, some of which preyed on rats
3 increases level of CO_2 in air
4 a) convert N_2 gas in air to nitrates
b) convert N-containing compounds from organic remains into nitrates
c) break down nitrates returning nitrogen gas to air

Food chains and energy flow

A four
A only some biomass/energy is transferred at each level/some lost in waste
A less loss of biomass/energy between each stage
1 a) used in running chemical processes
b) about 3 kJ, depending on loss in waste substances
2 select scale eg 1 cm = 10 000 kJ/m²/year; first trophic level = 88 000/10 000 or 8.8 cm, second = 1.4 cm, third is 0.16 cm and top trophic level is 0.01 cm

Atomic structure

A protons and neutrons in the nucleus; electrons circulating around the nucleus
A because number of positive protons equals number of negative electrons
A the mass number is the number of protons + the number of neutrons
1 a) neutron b) electron c) electron d) proton
2 the circulating electrons are negative and thus attracted to the positive nucleus
3 a) 8p 8n, 8e
b) 13p, 14n, 13e
c) 92p, 143n, 92e
d) 17p, 18n, 17e
e) 29p, 35n, 29e
4 atomic number is number of protons in an atom; mass number is total of protons and neutrons

Isotopes and mass

A isotopes have same number of protons, number of electrons, atomic number, chemical properties; they have different numbers of neutrons, mass numbers, physical properties
A the element contains a mixture of isotopes, relative atomic mass takes into account the proportions of each isotope
1 all have 8 protons but 8, 9, 10 neutrons respectively
2 relative atomic mass = (90 x 20) + (10 x 22) / 100 = 20.2
3 a) 1 + 1 + 16 = 18
b) 12 + 16 = 28
c) 40 + 12 + 16 + 16 + 16 = 100

Electronic configuration

A 3 sub-levels of 8, 10, 14 = 32 electrons
A the arrangement of the electrons in an atom
A 1st sub-level of shell 4 then 2nd sub-level of shell 3
1 missing numbers are: sodium 11 12 11 2.8.1
magnesium 24 12 12 2.8.2
iron 56 26 26 2.8.14.2
beryllium 4 4 5 2.2

2
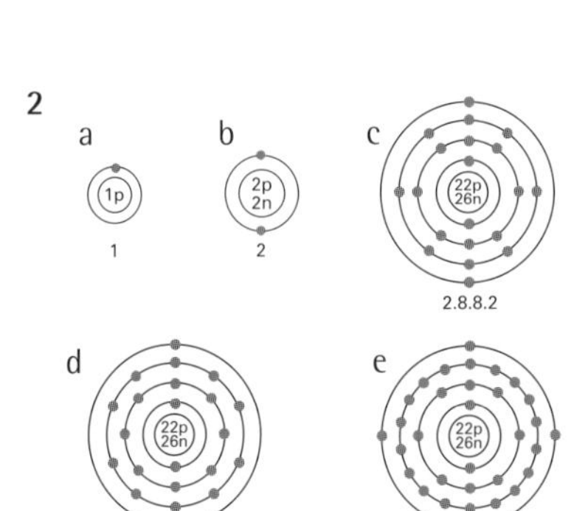

Ionic bonding

A a charged particle in which the number of protons does not equal the number of electrons
A there are two more positive protons than negative electrons
A the attraction between positive and negative ions
1 a) K^+ b) F^- c) Li^+ d) Al^{3+} e) S^{2-}
2 a) both have 2.8 electrons, OR both are positive, both have 12 electrons; sodium has a single positive charge but magnesium has a double positive charge sodium has 11 protons but magnesium has 12
b) both have 2.8 electrons, both have equal numbers of protons and neutrons, both have double charges; magnesium is positive but oxygen is negative, magnesium has 12 protons but oxygen has 8, magnesium has 12 neutrons but oxygen has 8
3
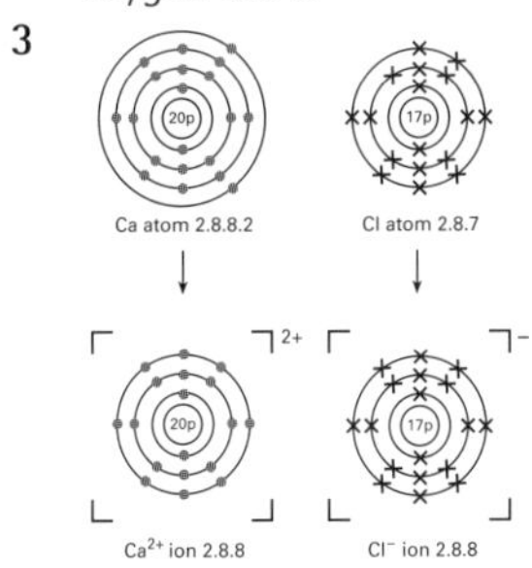

Covalent bonding

A a shared pair of electrons holding two atoms together, one electron from each atom makes up the pair
A two shared pairs of electrons holding two atoms together; hence a single bond is one shared pair of electrons
A there are no charges on the molecules
1
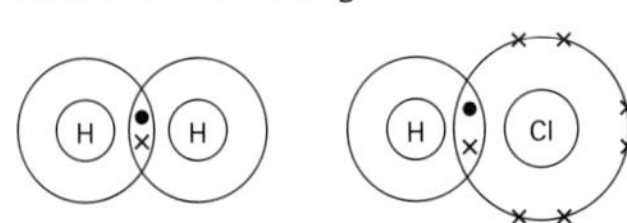

2 low melting and boiling points because inter-molecular forces are low; non-conductors because the molecules are electrically neutral

3 no single molecules but giant structures with strong covalent bonds which do not break down on heating

Chemical equations

A the quantities involved in the reaction

A Mg^{2+} O^{2-}; K^+ Cl^-

1 a) hydrogen + oxygen → water;
$2H_2 + O_2 \rightarrow 2H_2O$
b) zinc + steam → zinc oxide + hydrogen;
$Zn + H_2O \rightarrow ZnO + H_2$
c) nitrogen + hydrogen → ammonia;
$N_2 + 3H_2 \rightarrow 2NH_3$
d) sodium + water → sodium hydroxide + hydrogen;
$2Na + 2H_2O \rightarrow 2NaOH + H_2$

2 a) i) $4\,K + O_2 \rightarrow 2K_2O$
ii) $4\,K + O_2 \rightarrow 4K^+ + 2O^-$
b) i) $2NaOH + MgCl_2 \rightarrow Mg(OH)_2 + 2NaCl$
ii) $2Na^+ + 2OH^- + Mg^{2+} + 2Cl^- \rightarrow Mg^{2+} + 2OH^- + 2Na^+ + 2Cl^-$

Calculations from equations

A 12 + 16 + 16 = 44

A $2 \times 6.02 \times 10^{23} = 1.204 \times 10^{24}$

A 4 x (23 + 16 + 1) = 160 g

1 $2H_2O \rightarrow 2H_2 + O_2$
36g → 4g
36/4g → 1 g
9 g → 1 g therefore 9g of water is needed

2 $CuCO_3 \rightarrow CuO + CO_2$
124 g → 80 g
1g → 80/124 g
6.2 g → 80/124 x6.2 g therefore 4g of copper oxide is produced

3 $4Na + O_2 \rightarrow 2Na_2O$
92 g → 124 g
92/124 g → 1 g
92/124 x 1.24 → 1.24 g
0.92 g → 1.24 g therefore 0.92 g of sodium is required

Volume and formulae

A a) $3 \times 24\ dm^3 = 72\ dm^3$
b) $5 \times 24\ dm^3 = 120\ dm^3$

A a pure element always contains the same elements in the same fixed proportions by weight

1 a) ratio of nitrogen: oxygen = 7:16
divide each by its Ar = 7/14:16/16
= 0.5:1 = 1:2
Therefore formula is NO^2
b) ratio of sodium: oxygen = 4.6:1.6
divide each by its Ar = 4.6/23:1.6/16
= 0.2:0.1 = 2:1
Therefore formula is Na_2O

Acids, bases and neutralisation

A a compound producing H^+ ions in solution in water

A a compound producing OH^- ions in solution in water

A a reaction in which H^+ ions from an acid and OH^- ions from an alkali combine to form water molecules $H^+ (aq) + OH^- (aq) \rightarrow H_2O\ (l)$

1 a) produce high concentrations of H^+ ions; pH range 0, 1, 2; hydrochloric / nitric / sulphuric acids
b) produce high concentrations of OH^- ions; pH range 12, 13, 14; sodium/potassium hydroxide
c) produce low concentrations of H^+ ions; pH range 4, 5, 6; carbonic acid in the example in the text – there are others, e.g. ethanoic, lactic acid etc.
d) produce low concentrations of OH^- ions; pH range 8, 9, 10; ammonium hydroxide is the example in the text

2 a) i) $KOH + HNO_3 \rightarrow KNO_3 + H_2O$
ii) $OH^- + H^+ \rightarrow H_2O$

Redox and decomposition

A breaking up a compound by heat

A gain of oxygen; loss of hydrogen; loss of electrons

1 methane and carbon monoxide are oxidised, and chlorine and copper oxide are reduced.

2 a) $2Mg - 4e^- \rightarrow 2Mg^{2+}$: oxidation
$O_2 + 4e^- \rightarrow 2O^{2-}$: reduction
b) oxidising agent is oxygen; reducing agent is magnesium

Hydrocarbons

A simple organic compound that only contain hydrogen and carbon

A C_7H_{16}

```
  H H H H H H H
  | | | | | | |
H-C-C-C-C-C-C-C-H
  | | | | | | |
  H H H H H H H
```

A saturated hydrocarbons contain only single bonds; unsaturated hydrocarbons contain at least one double bond

A C_5H_{10}

```
  H H H H   H
  | | | |  /
H-C-C-C-C=C
  | | |    \
  H H H     H
```

1

```
  H H          H H H          H H H
  | |          | | |          | | |
H-C-C-H      H-C-C=C-H      H-C-C-C-Cl
  | |          |              | | |
  H H          H              H H H
```

2 a) reaction in which one atom in a molecule is directly replaced by another atom – the example in the text is the formation of chloroethane but there are many other examples

b) one in which a small molecule adds on across the double bond of an unsaturated molecule – the example in the text is the addition of H_2 to C_2H_4 but there are many other examples

3

```
  C4H10        + Cl2   →     C4H9Cl        + HCl
  H H H H                    H H H H
  | | | |                    | | | |
H-C-C-C-C-H    + Cl-Cl →   H-C-C-C-C-Cl    + H-Cl
  | | | |                    | | | |
  H H H H                    H H H H

  C3H6         + H2    →     C3H8
  H H H                      H H H
  | | |                      | | |
H-C-C=C-H      + H-H   →   H-C-C-C-H
  |                          | | |
  H                          H H H
```

Crude oil, petrol and polythene

A the separation of crude oil into its different fractions by heating and condensation

A breaking long chain hydrocarbons into short chain ones; short chain are more useful and in greater demand

A joining small, separate molecules (monomers) together into single long chain molecules (polymers)

1 a) $2\,C_2H_6 + 7\,O_2 \rightarrow 4\,CO_2 + 6H_2O$
b) $C_3H_8 + 5\,O_2 \rightarrow 3\,CO_2 + 4\,H_2O$

2 $n\,CH_2 = CHCl \rightarrow \{CH_2 - CH_2\}n$

Reactivity of metals

A a list of metals arranged in order of reactivity

A zinc is more reactive than copper, less reactive than magnesium

A rocks composed mainly of the compounds of one metal from which the metal can be extracted

1 It is a very reactive metal with stable compounds.

2 a) $Mg + ZnSO_4 \rightarrow MgSO_4 + Zn$
b) no reaction; copper is less reactive than magnesium
c) $Zn + CuO \rightarrow ZnO + Cu$
d) $Mg + Cu\ (NO_3)_2 \rightarrow Mg\ (NO_3)_2 + Cu$

3 a) $2\ HgO \rightarrow 2\ Hg + O_2$
b) zinc is more reactive than Hg and its compounds are thus more stable than mercury compounds

Electrolysis

A because the ions in molten NaCl are free to move but are fixed in place in solid NaCl.

A it acts as a solvent

A 98% pure copper obtained by extraction from copper ore

1 a) $2\ I^- - 2e^- \rightarrow I_2\ (s)$
b) $2\ O^{2-} - 4\ e^- \rightarrow O_2\ (g)$
c) $2\ K^+ + 2\ e^- \rightarrow 2\ K\ (s)$

2 The basic information required is:
ions present: $MgCl_2 \rightarrow Mg^{2+} + 2Cl^-$
at anode (+): $2Cl^- - 2e^- \rightarrow Cl_2$ (g)
at cathode (–): $Mg^{2+} + 2e^- \rightarrow Mg$ (s)

3 Because the basic electrode reactions are:
anode (+): $3O^{2-} - 6e^- \rightarrow 3O$ (oxidation)
cathode (–): $2Al^{3+} + 6e^-$
$\rightarrow 2Al$: (reduction)

Ammonia and fertilisers

A ammonia is used in fertiliser manufacture

A fertilisers containing nitrogen

A natural fertilisers are products of nature such as manure; artificial fertilisers are manufactured chemically

A fertilisers being washed by rain into lakes and rivers

1 $3H_2$ (g) + N_2 (g) $\rightarrow$ $2NH_3$(g)
(using 250atm / 550°C / Fe catalyst)

2 provide nutrients; produce high yields of healthy, large crops

3 Fertilisers washed into the lake encourage algae growth; cannot settle on the surface of flowing river water

4 algae dies; bacteria feed on this and multiply; increasing numbers of bacteria consume increasing amounts of oxygen; other pond life dies through lack of oxygen

The atmosphere

A released by photosynthesis of the first plant life

A to maintain balance

1 a) condensation of water vapour from the atmosphere
b) oxygen is released into the atmosphere from photosynthesis in plants

2 The formation of an ozone layer filtered out harmful UV radiation from the sun allowing complex living things – animal life – to develop

3 Two of: respiration by plants and animals; bacteria feeding on dead matter; burning fossil fuels

4 an open-ended question – answers might include: polar ice melting, further encroachment of deserts; more extreme weather patterns; effects on leisure industry; forced migration of population; effects on agriculture and food production, etc.

Rocks

A core, mantle and crust

A Earth was formed by solidification of molten rock and hence only igneous rock at first

A existing mountains are worn away by weathering / erosion; new mountains are formed by movements of the earth's crust

1 basalt – extrusive igneous rock – small crystals formed by rapid cooling on surface; granite – intrusive igneous rock – large crystals formed by slow cooling below surface

2 knowledge of the age/period of the fossil and the layer in which it is found provides information on the age of the layer and the sedimentary rock

3 constructive – plates move apart; sea-bed widens; magma rises to fill gap
destructive – plates collide; ocean plate slides under continental plate
conservative – plates slide past each other

The periodic table

A a table listing elements in order of atomic number and having elements which are chemically similar in the same vertical groups

1 three of A, B, C, D

2 A

3 three of: E, F, G, H

4 G

5 G

6 both are transition metals

7 i) either box below G in group VII
ii) top box of group I
iii) in either of the two boxes any box above H in group 0

8 J in the top box of group IV; K in the bottom box of group IV

9 the first box to the right of D; transition metal

10 i) box below E in group III
ii) top box of group 0
iii) in either of the two empty boxes in group II
iv) in either the box above or the box below F in group VI

Group I - the alkali metals

A inter-atomic forces are weaker in the larger atoms lower down the group

A ion formed during a reaction where an atom loses 1 electron

A because they all have one electron in their outer shell, so react similarly

A the ionic lattices can be broken down in water

1 acceptable ranges for the answers are:
a) 2.0–2.5g/cm^3
b) 12–20°C
c) 550–650°C

2 a) $2K + Br_2 \rightarrow 2KBr$;
$2K + Br_2 \rightarrow 2K^+ + 2Br^-$
b) $2Li + 2H_2O \rightarrow 2LiOH + H_2$;
$2Li + 2H_2O \rightarrow 2Li^+ + 2OH^- + H_2$
c) $4Na + O_2 \rightarrow 2Na_2O$;
$4Na + O_2 \rightarrow 4Na^+ + 2O^{2-}$

3 Rb; the one electron lost is lost easier and quicker from the Rb atom because it is further from the nucleus – attraction to the nucleus less, shielding greater – than it is in Li.

4 Na; sodium reacts by loss of one electron and this is easier than reaction by loss of two electrons as magnesium

5 a) Rb_2SO_4; $(Rb^+)_2SO_4{}^{2-}$
b) $NaNO_3$; Na^+ $NO_3{}^-$
c) KI; K^+I^-

First transition metals

A because the spaces in their penultimate shell allow electron movement

A because they are less strongly basic

1 zinc has a complete penultimate shell

2 atom gets smaller left to right, hence can pack closer and increase density

3 they react by electron loss to form positive ions

4 number of electron shells in use is same across the series

5 can lose electrons from the incomplete penultimate shell as well as the two outer electrons

Group VII - the halogens

A molecule containing just two atoms joined covalently

A chlorine – the electron gained goes to a shell nearer to the nucleus – more attraction to nucleus, less shielding – and it is therefore gained more easily

1 a) $Cl_2 + 2KBr \rightarrow 2KCl + Br_2$;
$Cl_2 + 2Br^- \rightarrow 2Cl^- + Br_2$
b) $I_2 + NaCl \rightarrow NaI + Cl_2$
$2Na + I_2 \rightarrow 2Na^+ + 2I^-$
c) $Br_2 + 2KI \rightarrow 2KBr + I_2$;
$Br_2 + 2I^- \rightarrow 2Br^- + I_2$

2 halogens react by gaining one electron; the nearer to the nucleus this goes the more attraction to the nucleus and the less shielding there is; hence the faster the reaction; thus top of group is fastest and bottom of group slowest
alkali metals react by loss of one electron; this is reverse of the above and so the reverse obtains; reaction is fastest at bottom and slowest at top of group

3 chlorine; gaining one electron is easier than gaining two

Group 0 – the noble gases

A when atoms exist singly and not joined to other atoms; noble gases are monatomic because complete outer shell does not necessitate joining to other atoms

1 the atoms are getting heavier and the inter-atomic forces are increasing

2 because they have a full outer shell of electrons

Rates of reaction

A particles gain energy and move faster

A increased pressure pushes particles closer; hence concentration increases

A reaction with powdered marble is very fast; CO_2 is released very quickly; mixture fizzes up the tube

A catalysts increase the rates of chemical reactions but are not themselves used up during the reaction

A a protein which catalyses reactions in the body

1 particles must collide with enough energy to reach or exceed the activation energy for the reaction

2 a) increasing temperature gives particles increased energy and speed; hence more fruitful collisions; hence increased rate
b) larger surface area in one reactant exposes more molecules to the other reactant; hence more fruitful collisions; hence increased rate

3 by increasing activation energy for the reaction

Energy in reactions

A exothermic – energy/heat given out
endothermic – energy/heat taken in

A change of energy/heat content of a material or system

A amount of energy required (+ kJ/mol) to break the bond

1 reactants

Reversible reactions

A a reaction in which products can react together to give the original reactants back again; equilibrium occurs when forward and backward reaction are at the same rate and there is no further change

A when conditions are changed in a system at equilibrium, the equilibrium will move to minimise the change

A if pressure increases, the system will minimise this effect by moving to fewer molecules and lower volume – i.e. move to the right.

1 a) system will move to absorb the extra heat; moves to left
b) system will absorb the pressure by moving to smallest side where there are fewer molecules; moves to right
c) catalyst has no effect on the equilibrium
d) system will move to replace D; moves to right
e) system will move to replace B; moves to left

Current, voltage and resistance

A $I = Q/t = 180\ C/360\ s = 0.5\ A$

A

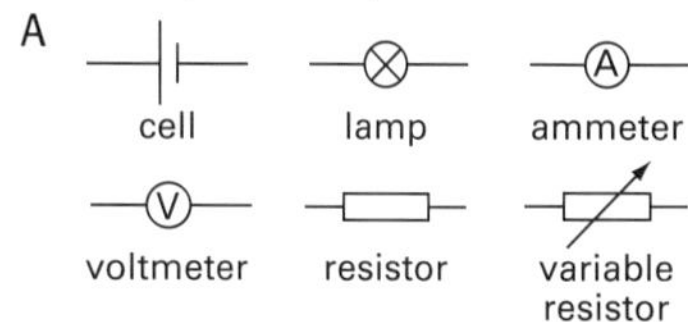

A $V = I\ R = 2\ A \times 10\ \Omega = 20\ V$

1 (a) $Q = I\ t = 4\ A \times (5 \times 60)\ s = 1200\ C$
(b) $E = V\ Q = 24\ V \times 1200\ C = 28800\ J$

2 $V = I\ R = 2\ A \times 5\ \Omega = 10\ V$

Electrical components 1

A V = IR

A The resistance of a wire increases as it gets hotter.

A Because it only allows current to pass in one direction.

1

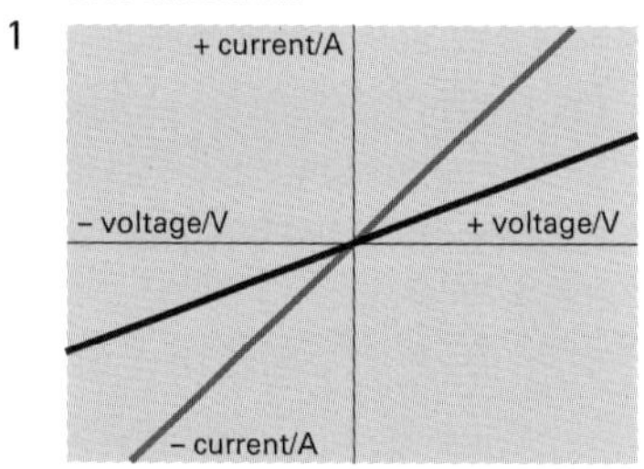

2 lamp A lights, lamp B doesn't light because the diode doesn't allow current to pass through that branch of the circuit.

Electrical components 2

A

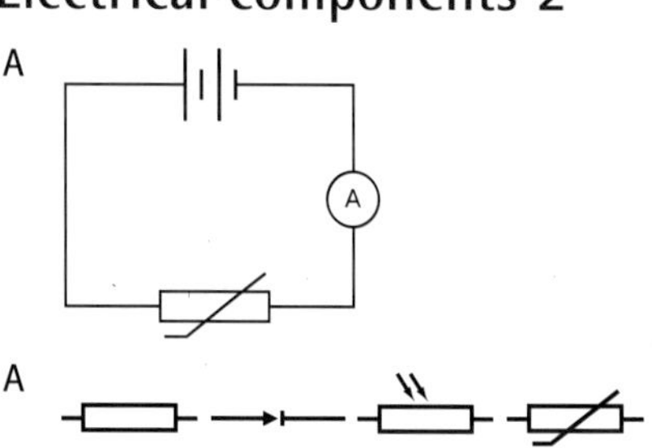

A

resistor diode LDR thermistor

1

resistance
time

2

A

Electricity, energy and power

A 3000 W = 3 kW

A energy = power × time = 60 W × 600 s = 36000 J

A 2 kW × 0.1 h = 0.2 kWh

A efficiency = $\frac{1.5\ kW}{3\ kW} \times 100\% = 50\%$

1 energy = V I t = 12 V × 3 A × 600 s = 21600 J

2 efficiency = $\frac{6\ W}{15\ W} \times 100\% = 40\%$

3 Energy input = V I t
= 12 V × 3 A × 300 s = 10800 J.
The motor is 75% efficient, so $\frac{3}{4}$ of this energy is useful.
$\frac{3}{4} \times 10800\ J = 8100\ J$

Electricity at home

A hertz – Hz

A blue: neutral, green/yellow: earth, brown: live

A If the fuse rating is much higher than the current needed by the appliance a fault may cause too much current to flow before the fuse breaks the circuit.

1 (a) conducting core needs to be a good conductor – such as copper
(b) outer covering needs to be a good insulator and flexible – such as a flexible plastic
(c) the fuse in a plug needs to melt before any of the other conductors – so needs a low melting point
(d) the casing of the plug needs to be a good insulator and rigid, so needs a rigid plastic

2 A 13A fuse should be used – a 3A fuse would break in normal use.

3 Direct current always flows in the same direction. Alternating current changes direction and size regularly.

Electric charge

A The hair stands on end because all the hairs have the same electric charge and so repel each other – getting as far apart as possible. The crackle happens when the charges discharge, making small sparks.

A The TV picture is made by an electron beam inside the television tube. The electrons bombarding the screen cause a charge to build up on the glass. This charge attracts charged particles of dust, which stick to the glass.

1 The jumper rubs the hair and charges it. The charge flows through the air as sparks.

2 When Julie walks across the carpet she becomes charged. The metal rail is a good conductor, so when she touches it the charge from here hand jumps across the gap, giving her a shock.

3 $I = Q/t = \frac{0.00012\ C}{1200\ s} = 0.1\ \mu A$

Magnets and electromagnetism

A The magnetic field due to an electric current in a coil can be made stronger by increasing the current, increasing the number of turns in the same length and by putting iron through the coil.

A The catapult force can be made larger by increasing the current in the wire or by increasing the strength of the magnetic field.

A The current is moving in opposite directions on each side, so the force is in the opposite direction.

1 More current through the coil, more turns of wire on the coil or a stronger magnetic field.

2 Reverse the current or reverse the magnetic field.

Electromagnetic induction

A A voltage is only induced when there is relative movement between the magnetic field and the wire.

A The output voltage would be greater and change at a greater frequency.

1 a) flick to left; (b) flick to left (c) flick to left; (d) needle stationary (e) flick to right.

2 The spinning magnet makes the magnetic field inside the coil change direction, which has the same effect as the magnetic field getting larger and smaller when the magnet is pushed in and out of the coil.

Transformers and power stations

A An alternating current in the primary coil makes a changing magnetic field in the iron core, which is needed to induce a voltage in the secondary coil.

A Mains electricity has to be generated as alternating current so that transformers can be used to step up the voltage for transmission to reduce energy losses.

1 Step-down transformer. The rails of the train set carry the current – so a low voltage is much safer.

2 a) $\frac{V_P}{V_S} = \frac{N_P}{N_S}$

$\frac{230\ V}{12\ V} = \frac{460 \times 12\ V}{\text{time taken}}$

$N_S = \frac{460 \times 12\ V}{230\ V} = 24$

b) $V_P I_P = V_S I_S$

$230\ V \times I_P = 12\ V \times 0.5\ A$

$I_P = \frac{6\ W}{230\ V} = 0.026\ A$

3 The lost energy warms the surroundings of the power station – as the fuel is burned, through friction in the turbines and generators, and the electric current generated will warm the wires in the generators and transformers.

Distance, speed, acceleration

A distance = speed × time
= 30 m/s × (20 × 60) s = 36000 m

A Velocity tells us the direction something is moving as well as the rate at which it is moving. Speed only tells us the rate at which it is moving.

A $\text{acceleration} = \frac{\text{change in speed}}{\text{time taken}}$

$= \frac{8400\ m/s}{480\ s} = 17.5\ m/s^2$

1 a) $\text{average speed} = \frac{\text{distance travelled}}{\text{time taken}}$

$= \frac{400\ m}{80\ s} = 5\ m/s$

b) He must accelerate from rest at the start, so if 5 m/s is his average speed, he must have moved faster than this at some stage.

2 $\text{acceleration} = \frac{\text{change in speed}}{\text{time taken}}$

$= \frac{30\ m/s}{10\ s} = 3\ m/s^2$

Forces

A The forces on an object moving at a steady speed are balanced – or there are no forces acting.

A Gravity acts down on the book. The table pushes up with a reaction force.

A The forces acting on an object that is accelerating are unbalanced. The net force is in the direction in which it is accelerating.

A $F = m\,a = 3000\ kg \times 3\ m/s^2 = 9000\ N$

1 a) $\text{acceleration} = \frac{\text{change in speed}}{\text{time taken}}$

$= \frac{15\ m/s}{10\ s} = 1.5\ m/s^2$

b) $F = m\,a = 1200\ kg \times 1.5\ m/s^2$
$= 1800\ N$

2 When Jane pushes Chris away an equal and opposite reaction force from Chris makes Jane move away.

Stopping forces

A There are resistive forces between the moving parts in a car. There are resistive forces between the tyres and the road – that enable it to move. There is also air resistance as the car moves through the air.

A tiredness, drugs, alcohol, distractions inside the car, poor visibility.

A the mass of the car; the speed the car was travelling before braking; the friction forces in the brakes; the road conditions

A the thinking distance and the stopping distance.

1 a) A heavily loaded car has a large mass so the deceleration will be less and it will travel further while braking- the stopping distance will be greater.
b) The car is moving quickly it has a greater change in speed to make so it will need to decelerate for longer and travel further – a greater braking distance. It will also travel further during the driver's reaction time, so the thinking distance will be greater. The stopping distance will be greater.
c) The friction between the tyres and a wet road will be less, the driver will have to apply the brakes gently to avoid skidding so the car will travel further while braking and the stopping distance will be longer.

2 a) distance = speed × time
= 30 m/s × 0.5 s = 15 m
b) total stopping distance
= thinking distance + braking distance
= 15 m + 64 m = 79 m

Falling

A You need to know your mass in kilograms. The force of gravity on you (your weight) is 10 newtons for each kilogram.

A Air resistance depends on the shape of the object and the speed at which it falls.

A The terminal velocity of an object depends on the weight of the object and its shape.

1 The drag chute increases the resistance and shortens the stopping distance – the aircraft carrier deck is much shorter than a runway.

2 On Earth air resistance acts on the hammer and feather. The terminal velocity of the feather is much less than for the hammer, so the hammer falls more quickly.

Properties of waves

A $v = f\lambda = 440\text{ Hz} \times 3.4\text{ m} = 1496\text{ m/s}$

A angle of incidence = angle of reflection

A Refraction occurs because change speed when they pass from one material to another.

1 Transverse waves: the oscillation is at right angles to the direction in which the energy travels
Longitudinal waves: the oscillation is in the same direction as the direction the energy is carried

2 $v = f\lambda = 440\text{ Hz} \times 0.75\text{ m} = 330\text{ m/s}$

3 Echoes are sounds reflected from hard surfaces.

Wave behaviour

A The light from the bird is refracted as it enters the water, so the bird appears to be further away than it is.

A Diffraction happens best when the wavelength is a similar size to the size of the gap.

1

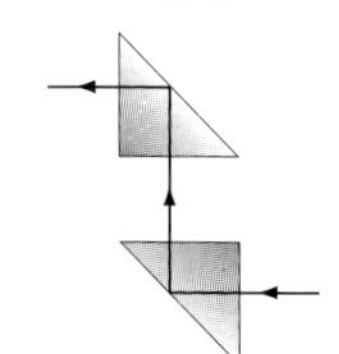

2 You can hear around corners – you do not need to be in direct line of sight with a sound source to hear it.

Electromagnetic spectrum

A gamma rays: killing cancer cells; x-rays: to produce images of human bones; uv: tanning skin; violet–red: visible light; infrared: toasters; microwaves: mobile phones; radio waves: radar guns

A X-rays can damage cells in the body by ionising molecule sin the cell. The greater the exposure to x-rays the greater the chances of lasting damage.

1 All electromagnetic waves travel at the same speed, they are all transverse and they can all travel through a vacuum.

2 a) x-rays used for imaging – a hazard because they damage cells in the body
b) microwaves used in communications and cooking; hazardous because they some wavelengths cause heating of water.
c) ultraviolet waves used to sterilise surgical equipment, hazardous because they can cause skin cancer.

Communicating with waves

A they can transmit coded information quickly over long distances

A Optical fibres can carry more information than an equivalent thickness of copper. They cannot easily be 'tapped'.

A Microwaves travel in straight lines, the Earth curves.

A Digital signals can carry more information and suffer less from interference.

1 Optical fibres can carry more signals than copper wire; the fibres are cheaper than copper; the signal does not escape from the cable, so cannot be detected; the energy loss in the fibre is less than in copper; the signal suffers less from interference.

2 Analogue signals can have any value, related to the size of the original sound. A digital signal is a string of 0s and 1s which are binary numbers to show how the voltage of the signal is changing.

Sound and ultrasound

A Sound waves are longitudinal waves

A X-rays damage cells in the body – there are no known side effects to using ultrasound.

1 Sound travels 60 m (there and back) in 0.2 s.
$\text{speed} = \frac{\text{distance}}{\text{time}} = \frac{60\text{ m}}{0.2\text{ s}} = 300\text{ m/s}$

2 Ultrasound is reflected at the boundary between different materials.

Structure of the Earth

A Plates move apart, allowing molten rock to come to the surface. Plates move together, forming mountains by folding; causing volcanoes to erupt or earthquakes. Plates slide past each other causing earthquakes.

A The Atlantic ridge is formed at the junction between the South American and North American plates and the African and Eurasian plates.

A Primary P waves and Secondary (S) waves.

1 P waves are longitudinal waves and can pass through solids, liquids and gases; S waves are transverse and travel more slowly. They can only pass through solids.

2 P waves arrive first. Fast P waves travel at 13 km/s. So there will be a time delay of 230 s.

The Earth in space

A Mercury, Venus, Earth, Mars, Jupiter, Saturn, Uranus, Neptune, Pluto

A Its orbit is much more elliptical and at an angle to the plane of the other planets

1 Between Saturn and Uranus. Between 30 years and 80 years.

2 The further a planet is from the Sun, the longer it takes to complete one orbit.

Moving in orbit

A The pull of gravity between two objects depends on the mass of the two objects and how far apart they are.

A Telephone transmissions, television transmissions, weather monitoring, environmental monitoring, spying, astronomical observations;

1 $W = mg$
a) On Earth, $W = 50\text{ kg} \times 10\text{ N/kg}$
$= 500\text{ N}$
b) On the Moon, $W = 50\text{ kg} \times 1.6\text{ N/kg}$
$= 80\text{ N}$

2 The asteroid is very small compared to the Earth, so the mass will be small and its gravitational pull will be small – it will be much easier to leave the surface.

Our place in the Universe

A

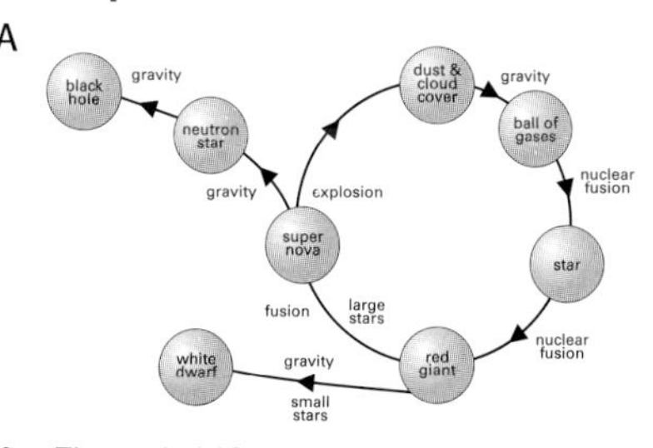

A The red shift suggests that the Universe is expanding.

A Only you can answer this – you might be happy to accept the evidence of radio signals that seemed to make a patter to communicate – or you might want to see the living things with your own eyes – or something in between.

1 Our Sun will probably become a white dwarf, gradually fading as it cools down.

2 When we look at stars we are looking at light that was generated a long time ago – the light from distant stars took thousands of years to reach us.

3 distance = speed × time
= 300 000 km/s × 480 s
= 144 million km

Work, power and energy

A work done = force × distance
= 100 N × 1.5 m = 1500 J

A from the food he eats

A potential energy = m g h
= 70 kg × 10 N/kg × 3 m = 2100 J

A The energy shakes the molecules in the ground and the cans so they get a little warmer; some energy is carried away by sound.

1 potential energy = m g h
= 30 kg × 10 N/kg ×5 m = 1500 J

2 a) work done = force × distance
= 700 N × 3 m = 2100 J
b) power = $\frac{\text{work done}}{\text{time}} = \frac{21000}{30\text{ s}}$ = 70 W
c) The motor also has to lift the chair and itself.

Energy transfers

A Atoms move more quickly.

A The electrons transfer energy through the metal.

A Convection currents carry the warm air up the staircase.

A When the sweat evaporates it carries away energy from the surface of your skin.

A Frying in a pan uses conduction to transfer energy from the cooker to the food ; cooking under the grill uses radiation; cooking soup in a saucepan uses convection for the energy from the bottom of the pan to be transferred through the soup.

1 fibre glass, loose fill polystyrene

2 Energy is transferred round the room by convection currents rather than by radiation – the radiators are often painted a light colour, which makes them poor radiators.

3 Energy is lost through the roof by convection and conduction. A layer of insulating material creates air pockets, which are poor conductors and prevent convection.
Energy is lost through draughts through gaps between doors and windows and their frames. Draught proofing materials can be stuck to the frame to make a better seal.
Energy is lost by conduction through the glass of windows; double-glazing inserts an insulating layer of air between to layers of glass.

Energy resources

A All the energy eventually warms up the surroundings – through friction, braking and air resistance

A Generating electricity by fossil fuels causes increase in acid rain and global warming; using fossil fuels for electricity generation means they are not available for other uses – such as making materials; the fossil fuel supply is limited and will run out;

A Insulate loft, reduce draughts, fit double glazing.

1 fossil fuels are used to make plastics

2 only 35% of the energy from the coal becomes a useful output as electricity.

3 A bus carries 30 – 80 people, so reducing the number of cars on the road, less fuel is used.

4 Advantages of wind and water are that they are renewable resources that will not run out and they are non polluting, and do not increase global warming. Disadvantage of wind power is that it is only available when the wind blows; water power depends on a good water supply, so depends on being in the right place.

What is radioactivity?

A protons and neutrons

A they ionise the air.

A negative charge

A lead stops gamma radiation

1 alpha α helium nucleus – 2 protons and 2 neutrons few centimetres paper
beta β electron few metres few millimetres of aluminium
gamma γ electromagnetic radiation very long way thick piece of lead or thicker concrete

Radioactive decay

A do not eat sea food; move to a non granite part of the country, not fly in aeroplanes

A Half life is the time for half the radioactive nuclei to decay

A Radiocarbon dating is for things made from once living material

1 a) 6 days is three half lives
b) The amount will have halved three times:
12 mg → 6 mg → 3 mg → 1.5 mg

2 It loses three quarters which means one quarter is left. In 1 half life it falls to half, in second half lifeit falls to one quarter – so two half lives, 110 s.

Using radioactivity

A The fact that radiation ionises is used in detection, its properties of penetration help identify the type of radiation.

A Gamma

A Gamma penetrates paper easily and there would be no detectable change in intensity after passing through paper.

1 Radiation ionises molecules in the cells of the body – protection limits the exposure to radiation. The further source is away, the less radiation will reach you; shielding reduces the radiation reaching the body. The shorter the exposure time the less risk of permanent damage.

2 Alpha emitters have such a low penetration, they would damage the cells close to the source, rather than the cancerous cells.

3 Gamma is used because it can penetrate the ground and so be detected at the surface.

Glossary

A

acceleration Rate at which a moving object is speeding up or slowing down. Measured in metres/second^2 (m/s^2).
acid Compound which produces H^+ ions in aqueous solution; pH below 7.
activation energy Energy required for reaction to occur when particles collide.
addition reaction Reaction in which a small molecule adds on across a double bond.
ADH Antidiuretic hormone which controls the production of urine in the kidneys.
aerobic respiration Oxygen is used in transferring energy from glucose in cells.
alkali Compound which produces hydroxide (OH^-) ions in aqueous solution; pH above 7.
alkali metals Elements in group I with one electron in the outer shell.
alkenes Unsaturated hydrocarbons with a double bond between the carbon atoms.
allele Version of a gene for a characteristic.
alternating current (a.c.) Electric current which regularly changes size and direction.
amp Electric current is measured in amps. 1 amp = 1 coulomb/second.
amplitude Distance from crest of a wave to the place where there is no displacement.
anaerobic respiration Energy is transferred from glucose without oxygen.
analogue signals Carries information by copying the changing pattern of the waves in the original information.
angle of incidence The angle between the direction of a wave and the normal.
artery A blood vessel with a thick muscular wall and a narrow space inside.
atom All elements are made of atoms. Consists of a nucleus containing protons and neutrons surrounded by electrons.
atomic number (Z) Number of protons in the nucleus of the atom.
auxins Hormones controlling plant growth, found in tips of growing shoots and roots.
Avogadro number Atoms/molecules in 1 mole of an element/compound; 6.02×10^{23}.

B

background radiation Radioactive materials (in rocks, air, particles from outer space) are sources of background radiation.
base Metal hydroxides, oxides, carbonates. Alkalis are bases that dissolve in water.
bond energy Energy required to break the bond. Units: kJ / mol.

C

capillary A thin-walled narrow blood vessel.
carbohydrates Foods made of carbon, hydrogen and oxygen. Eaten for energy.
carbon cycle Shows how the carbon is recycled via carbon dioxide in atmosphere.
carnivore Animal which eats other animals.
catalyst Changes the rate of a chemical reaction without being changed itself.
cell All living things are made of cells – the smallest units of living matter. All cells have a nucleus, cytoplasm and surface membrane.
chlorophyll Causes the green colour in plants, necessary for photosynthesis.
chloroplast Plant cells have chloroplasts that carry out photosynthesis using chlorophyll.
chromosome The nucleus contains chromosome pairs which contain many genes.
cloning A process of making an identical copy of a living thing.
combustion The reaction of a substance burnt in oxygen. It is an oxidation reaction.
compound Elements join together to form compounds by forming bonds.
concentration How much of a substance is dissolved in water, measured in moles per litre or moles per dm^3.
condensation Molecules in a vapour return to a liquid, losing energy as they do.
conductor An electrical conductor is a material which allows an electrical current to pass easily. It has a low resistance. A thermal conductor allows thermal energy to be transferred through it easily.
cornea Transparent outer coating of the eye.
coulomb (C) Measures electric charge.
covalent bond Bond between atoms forms when atoms share electrons to achieve a full outer shell of electrons.
critical angle Angle of incidence that results in waves refracting through 90°.
cytoplasm The living contents of the cell where the chemical processes take place.

D

decomposition A reaction in which substances are broken down, by heat, electrolysis or a catalyst.
diffraction This happens when waves pass through a gap which is comparable in size to the wavelength of the wave.
diffusion The effect of randomly moving particles in a liquid or gas, gradually spreading from an area of high concentration to one of low concentration.
digestion Food passes through the gut and is broken down into small enough molecules to pass into the blood. Undigested food passes out of the anus as solid waste.
digital signal Carries information as a series of 1s and 0s which code the values of the original signal.
diode Electrical component that only allows electric current to pass in one direction.
direct current (d.c.) Flows in same direction.
distillation Separation of mixture into one or more components by evaporation and condensation of the components with the lower boiling points.
DNA Chemical which makes chromosomes.
double bond Two shared pairs of electrons between two atoms.

E

effector Part of the body which responds to a stimulus, such as a muscle or gland.
efficiency A measure of how effectively energy is transferred in a system.
electric charge Electrons carry a negative charge, protons a positive one. Attract and repel each other. Measured in coulombs (C).
electric current Flow of electric charge around a circuit. Measured in amps (A).
electrolysis When an electric current passes through a solution or molten solid.
electrolyte The liquid which conducts an electric current during electrolysis. It contains ions which carry the current.
electromagnetic induction When a conducting wire moves relative to a magnetic field, a voltage is induced across the wire.
electromagnetic spectrum Family of waves which travel at the same speed in a vacuum.
electron A very small negatively-charged particle, surrounding the nucleus found in an atom. In a neutral atom, the number of electrons around the nucleus equals the number of protons in the nucleus.
electronic configuration Describes the arrangement of electrons within the energy levels or shells around the nucleus.
element All atoms of an element have the same atomic number, the same number of protons and electrons and so the same chemical properties.
endothermic Reaction that absorbs energy.
enzymes Made of protein, they are biological catalysts. Molecules become temporarily attached to the enzyme during biological processes.
epicentre The place in the earth that an earthquake occurred. Seismic waves spread out from the epicentre.
equilibrium If the rates of forward reaction and back reaction in a reversible reaction are equal, the reaction is in equilibrium.
evaporation Molecules near the surface of a liquid may leave the liquid to become a vapour – this is called evaporation.
exhalation (expiration) The process of decreasing the space inside the chest so air pressure rises and air flows out of the lungs.
exothermic Reaction that gives out energy.

F

food chain Shows the feeding relationship between living things and the way energy and nutrients flow through the system.
food pyramid Gives information on different living organisms in a food chain.
food web Shows how food chains are interlinked and how organisms in an ecosystem depend on each other.
fossil fuel Created over millions of years by the decay and compression of living things.
fractional distillation A mixture of several substances (e.g. crude oil) are distilled; the evaporated components are collected as they condense at different temperatures.
frequency The number of waves produced each second, measured in Hertz (Hz).
fusion During nuclear fusion two small atomic nuclei collide together with such energy that they fuse, releasing energy as electromagnetic waves.

G

galaxy A collection of many stars held together by gravity.
gas molar volume Volume occupied by 1 mole of any gap at r.t.p.; 24 dm^3.
gene A short length of DNA which is the code for a protein.
genetic engineering Taking genes from one living thing and inserting them into the DNA of another to change its characteristics.
giant lattice Compounds formed by ionic bonds form a giant structure in a lattice.
giant structure Sometimes covalent bonds produce very strong giant structures.
glucose A simple carbohydrate produced by plants during photosynthesis. Some is used as an energy source by the plant, the rest is converted to starch and stored in the leaves.
gravitational potential energy When an object is lifted against the pull of gravity, energy is transferred to the gravitational field. This is gravitational potential energy.
gravity The force which pulls objects to the ground. Gravitational attraction acts between any two objects with mass. The strength of the gravitational field is measured in newtons/kilogram (N/kg).
groups The groups in the periodic table are those elements which have the same number of electrons in their outer shells, and so have similar chemical properties. A group lies in the same column in the table.

H

half-life The time it takes for half a quantity of radioactive material to decay to a new substance.
halogens Elements in group VII of periodic table with 7 electrons in the outer shell.
heat of reaction Heat change (ΔH) produced in a reaction.
herbivore Animal that only eats plant matter.
hertz Measurement of frequency.
hormone Chemicals which help to coordinate life processes.
hydrocarbons Group of compounds which contain the elements hydrogen and carbon.

I

igneous rock Forms when molten rock cools.
inhalation (inspiration) The process of increasing space inside the chest, so the air pressure drops and air flows into the lungs.
inherited disease A disease passed on from one generation to the next from the genes of the parents to the child.
insulator An electrical insulator does not allow an electrical current to pass – it has a very high resistance. A thermal insulator does not allow thermal energy to be transferred through it easily.
insulin Hormone involved in the control of sugar levels in blood.
ion When atoms lose an electron, they become a positive ion. When they gain an electron, they become a negative ion.
ionic bond Forms when an electron is transferred from one atom to the other, forming a positive-negative ion pair.
iris Opaque tissue which controls the size of the pupil and so controls the amount of light which can pass into the eye.
isotope Atoms of the same element with different numbers of neutrons.

J

joule energy is measured in joules.

K

kidney The organ which controls water and salt balance and the pH of the blood. Kidneys excrete urea, hormones and medicines in the urine they produce.
kinetic energy The energy of a moving object.

L

LDR (light-dependent resistor) Resistance value of the component depends on the level of light falling on the component.
Le Chatelier's principle If a change is applied to a system in equilibrium, the system will move to minimise the change.
longitudinal wave A wave whose oscillations are in the same direction as the energy is travelling.

M

macromolecule Giant structure in which atoms are covalently bonded to each other continuously.
magnetic field A magnetic field is the region in space around a magnet in which other magnets are affected.
mass A measure of the amount of material in an object. Measured in kilograms (kg).
mass number (A) The number of protons and neutrons in the nucleus of an atom.
meiosis A parent cell divides to create four new sex cells, each with a different combination of chromosomes.
metamorphic rocks Formed when rocks are changed by heat or pressure.
microbe Small single-cell organisms, such as bacteria.
mitosis The process of cell division which results in two identical cells.
mole The relative atomic/molecular mass of an element/compound in grams.
monomer A simple molecule.
motor neurone A nerve cell that transmits a signal from the spinal cord to the effector.
mutation Spontaneous change in a gene or chromosome which may result in a change in cells characterised by the gene.

N

negative ion (anion) Formed when an atom gains an electron to achieve a full outer shell of electrons.
nephron The structure within the kidney which filters out waste from the blood and allows useful materials to be reabsorbed.
neutralisation The reaction between an acid and a base; $H+ + OH^- -> H_2$
neutron A small, uncharged particle found in the nucleus of the atom.
newton Measurement of force in newtons (N). 1 newton is the unbalanced force needed to give a mass of 1 kg an acceleration of 1 m/s^2.
nitrogen cycle Shows how nitrogen is recycled so is available to plants and animals.
noble gases The elements in group 0/VIII of the periodic table, with a full outer shell of electrons and so are unreactive.
normal A line perpendicular to the surface at the point where a wave crosses the surface.
nucleus of a cell Made of genetic material (e.g. DNA), it contains instructions for the cell.
nucleus of an atom An atom's small, dense, positively-charged nucleus, where most of the mass is concentrated, containing protons and neutrons.

O

oestrogen A sex hormones in females which causes puberty.
ohm Ω Measurement of electrical resistance.
optic nerve Nerve fibres from the light-sensitive cells on the retina form the optic nerve passing from the eye to the brain.
optical fibres Very fine threads of glass through which light can pass, even when the fibre is curved around a corner.
ore Rock containing a mineral which can be extracted.

osmosis A special case of diffusion – a solvent passes through a membrane from a weak to a stronger solution. The solute cannot pass through the membrane.
oxidation A reaction in which oxygen combines with a substance; a reaction where electrons are removed; a reaction where hydrogen is removed from a substance. – gain of oxygen, loss of hydrogen, loss of electrons in a reaction
oxygen debt During vigorous exercise anaerobic respiration occurs in muscle tissue. Lactate is produced in the absence of oxygen. When exercise stops, oxygen is used to respire the lactate.

P

periodic table Lists the elements in order of atomic number. Arranged into rows called periods and columns called groups.
periods The periods of elements in the periodic table are the elements in which the same outer shell is being filled up. A period of elements in the periodic table all lie in the same row.
pH The pH scale is a measure of how acid or alkali a solution is. The pH scale runs from 1 to 14.
photosynthesis Plants process water and carbon dioxide to produce glucose by the process of photosynthesis.
plasma The yellow liquid part of blood which contains many dissolved substances.
polymer a large molecule formed from many monomers by polymerisation.
polymerisation The reaction in which many identical monomers are joined together to make a polymer.
positive ion (cation) Formed when an atom loses an electron to achieve a full outer shell of electrons.
power Rate at which energy is transferred in a system. Power is measured in watts (W) and kilowatts (kW).
progesterone A sex hormones which causes puberty in females.
proteins Foods made from carbon, hydrogen and nitrogen. Proteins are used to build enzymes and cytoplasm.
proton Small positive particle in the nucleus of the atom.
pupil The hole through which light passes into the eye.

R

radioactive decay Random process in which the nucleus of an atom becomes more stable by losing particles and energy.
reactivity series List of metals in order of decreasing reactivity. Any metal will displace a metal below it in the Reactivity Series from its solution.
receptor A part of the body which detects a stimulus.
red blood cells Contain haemoglobin which transports oxygen to cells and returns carbon dioxide to the lungs.
redox reaction Oxidation and reduction always take place together. The combined reaction is called a redox reaction.
reduction Reaction in which oxygen is removed from a substance; or a reaction where electrons are gained or a reaction where hydrogen is gained by a substance – loss of oxygen, gain of hydrogen, gain of electrons in a reaction.
reflection Reflection of waves happens when the waves bounce off a surface.
refraction Refraction of waves happens when waves change direction due to a change in speed.
relative atomic mass Mass number of an element taking into account the proportions of each isotope.
relative molecular mass Sum of the relative atomic masses of all the atoms in one molecule of a compound.
resistance A measure of how difficult it is for an electric current to pass. Measured in ohms (W).
respiration The process which transfers energy from food to the organism.
retina Light-sensitive layer at back of eye.
reversible reaction Chemical reactions which can go both ways. The direction depends on the condition of the reactants.

S

satellite A body in orbit around a planet. Moons are natural satellites. There are many man-made satellites in orbit.
saturated compound An organic compound containing only single bonds
sedimentary rocks Formed when rock fragments are deposited and pressed together.
seismic waves Carry energy away from an earthquake.
sensory neurone Nerve cell carrying a signal from a receptor.
shells of atoms Electrons are grouped within an atom in regions of space called shells or energy levels.
solar system The Sun, nine planets, the asteroid belt and a number of comets.
speed The rate at which something moves. Speed is measured in metres / second (m/s).
stomata Pores (openings) in the surface of the leaf, through which transpiration occurs.
substitution reaction Reaction in which one atom is directly replaced by another

T

temperature regulation The body controls its internal temperature by varying blood flow, sweating and shivering.
terminal velocity When an object moves through a fluid, there is resistance to its motion. The faster it travels, the greater the resistance. When the driving force is balanced by the resistive forces, the object is moving at a maximum speed called its terminal velocity.
testosterone The sex hormone which causes puberty in males.
thermistor An electrical component whose resistance changes with temperature.
total internal reflection occurs when the angle of incidence is greater than the critical angle.
transformer Pair of wire coils linked by a piece of iron. A changing current in one coil creates a changing magnetic field in the iron, which causes a changing voltage in the other coil by electromagnetic induction.
transition metals The group of elements in the periodic table between Group II and III.
transpiration the loss of water from a plant by evaporation from stomata.
transverse wave Oscillations are at right angles to the direction in which the energy is travelling.

U

unsaturated compound An organic compound containing at least one double bond.

V

vein A blood vessel with a thinner wall, with a larger space and lower pressure than an artery.
velocity Tells us the speed of a moving object and the direction it is moving.
villi The small intestine is lined with villi, small 'fingers' of gut which increase the surface area for digestion.
volt Electrical potential difference (voltage) is measured in volts. 1 volt = 1 joule / coulomb.
voltage Voltage (potential difference) between two points in a circuit measures the difference in energy carried by the electric charge passing through the circuit. Voltage is measured in volts (V).

W

watt Power is measured in watts (W). 1 watt is 1 joule / second.
wavelength The distance from the crest of one wave to that of the next.
weight The force of gravity on an object. Measured in newtons (N).
white blood cells Fight infection. They make antibodies and overcome bacteria.
work When a force moves an object. When work is done, energy is transferred. Measured in joules. When a force of 1 newton acts through 1 m in the direction of the force, 1 joule of work is done.

Last-minute learner

Cells

- The smallest unit of life is a cell. Groups of cells of the same type form tissues and tissues build organs with have specialised functions.
- Common cell structures include: **cell surface membrane** (controls what moves in and out of a cell), **cytoplasm** (where chemical reactions occur), **nucleus** (contains genetic material).

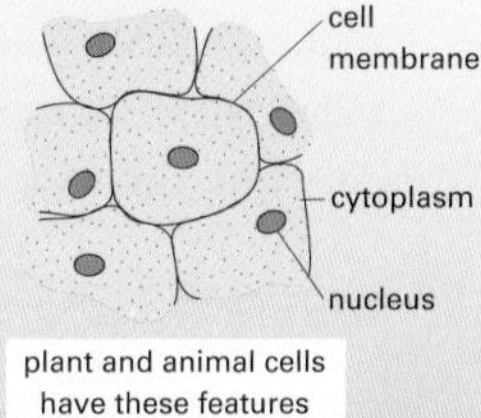

plant and animal cells have these features

- Plant cells have additional structures: **cell vacuole** (gives plants support and maintains shape), **chloroplasts**, containing chlorophyll (important for photosynthesis) and **cell walls** (give support).

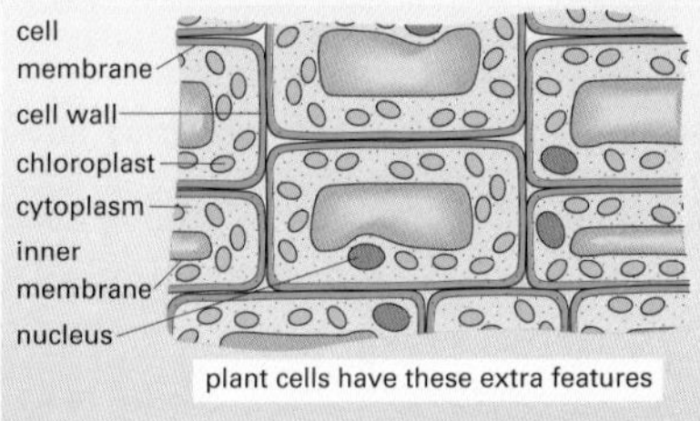

plant cells have these extra features

Life processes

- Life processes are the main activities carried out by body systems: **respiration, feeding, sensitivity, movement, reproduction, growth, excretion.**
- Nutrition comes from food, providing **energy** for processes and **raw materials** for new cells.

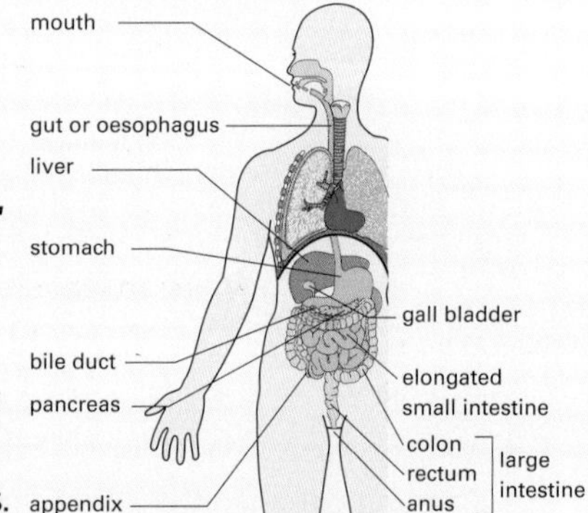

- A balanced diet includes: **carbohydrates, proteins, fats, water, minerals** and **vitamins** in the correct proportions and in levels appropriate to an individual's lifestyle.
- Food is broken down into smaller particles as it passes through the gut. Digested particles are absorbed into the bloodstream for transport to body tissues.
- Enzymes are mainly responsible for digestion: **carbohydrases** digest carbohydrates, such as starch; **proteases** digest proteins; **lipases** digest fats/lipids.
- The shape of an enzyme molecule is vital to its function. The active site is a part of the enzyme where molecules 'lock on' and react. Changes in pH, temperature or other conditions can destroy an enzyme by changing its shape.
- The heart is muscular and beats throughout life. At each beat the atria fill and contract, and then the ventricles fill and contract. The left and right sides beat simultaneously but the blood flow is separate.
- Arteries, veins and capillaries are the blood vessels that form the circulation to all cells. Blood contains dissolved substances, such as glucose and salts, plasma proteins (e.g. antibodies), platelets and red and white cells.
- Ill health may be caused by injury, inherited disease, or infection. Micro-organisms may cause infection.
- The body's main defences against infection include skin, cilia and white blood cells.
- Immunisation is a way of protecting ourselves against disease by causing an immune response.
- Lifestyle (diet, exercise, drug abuse, including smoking and drinking alcohol) has a significant effect on health.
- Breathing movements change the air pressure inside the lungs. During inhalation the pressure inside the lungs is less than ouside the body, and vice versa in exhalation.
- Respiration transfers energy from food to living cells. Aerobic respiration involves oxygen:
  $C_6H_6O_6 + O_2 \rightarrow CO_2 + H_2O$ (+ energy transferred)
  Anaerobic respiration happens in the absence of oxygen:
  $C_6H_6O_6 \rightarrow$ lactate (+ some energy transferred)
  $C_6H_6O_6 \rightarrow C_2H_5OH + CO_2$ (+ some energy transferred)

All about plants

- **Photosynthesis** takes place in the choroplasts, where light energy is absorbed by chlorophyll. Simple raw materials are converted into carbohydrate and oxygen.
  carbon dioxide + water $\xrightarrow[\text{sunlight}]{}$ glucose + oxygen
- The **rate of photosynthesis** is determined by the conditions: light intensity, concentration of carbon dioxide and a suitable temperature.
- **Osmosis** is the diffusion of water molecules, which follows a water concentration gradient.
- **Transpiration** is the loss of water from plant surfaces, which draws water up through the plant. Transpiration happens quickest in warm, dry and windy conditions.
- **Auxin** is a plant hormone, which causes uneven growth on either side of a shoot or root tip. In shoots it increases the rate of cell growth. In roots it slows the rate of cell growth.
- The growth of a plant towards light is **phototropism**. There is more auxin on the side of the shoot away from light, so it grows faster and curves towards light.
- Auxins are used to: make roots grow on shoot cuttings, kill weeds by abnormal growth, help fruit ripening

Regulating body processes

- A **synapse** is a microscopic gap between two neurones. **Neurotransmitters** are chemicals that diffuse across the synapse, initiating the **impulse** in the next neurone.

part of the nervous system	what it does
receptor cells e.g. skin, organ	detects a stimulus, such as a change in temperature
sensory neurone/nerve	carries impulses from receptors to central nervous system (CNS)
central nervous system (brain and spinal cord)	processes information detected by receptors/sense organs
motor neurone/nerve	carries impulses from CNS to effector, e.g. muscle or gland
effector tissue or organ	carry out action e.g. gland makes hormone, muscle moves

- **Hormones** are chemicals that help to **control conditions** within the body and coordinate life processes; they are carried in blood and act on target organs.
- **Menstrual cycle** = 28 days. **Ovulation** is on day 14, stimulated by **FSH. Fertilisation** happens in oviduct. **Implantation** happens in the uterus. **Oestrogen** causes womb lining to thicken – **progesterone** maintains it.
- **Fertility drugs** use hormones to stimulate egg/ovum production. The contraceptive pill may contain **oestrogen** and **progesterone** to prevent ovulation/fertilisation.
- Skin helps **thermoregulation** by adjusting blood circulation near body surface (**vasodilation** and **vasoconstriction**) and amount of sweat. It is waterproof, protecting the body from drying out and infection.

gland	hormone and result
pituitary	important in controlling growth rate; menstrual cycle; milk production; the thyroid
thyroid	thyroxine: regulates rate of chemical activities
pancreas	insulin: reduces blood sugar level glucagon: increases blood sugar level
adrenal	adrenaline: prepares body for rapid activity
testis	testosterone: causes secondary sexual characteristics and sperm production
ovary	oestrogen: causes secondary sexual characteristics and helps control menstruation

Variation, genetics and ecology

- Variation describes the differences between living things. It is caused **genetically**: through inherited genes and by chromosomes exchanging DNA during cell division, or **mutation** can occur by environmental factors.
- A **gene** is a short chunk of DNA, coding for a particular **protein**, giving rise to a **characteristic**.
- **Charles Darwin** and **Alfred Wallace** contributed to the theory of evolution. Mutations cause **genetic variation**, causing **new characteristics**, which may be **beneficial** (so more likely to be **passed on** to future generations) or **harmful** to an indivdual's survival.
- **Sexual reproduction** uses **meiosis**: two parents each provide a nucleus via a haploid sex cell, causing genetic variation in offspring. Plants and animals are bred selectively for desirable characteristics, over generations.
- **Asexual reproduction** uses **mitosis** and one parent, which is genetically identical to its offspring. **Cloning** is an example of asexual reproduction, e.g. plant cuttings and splitting embryos in cattle breeding.
- **Genetic engineering** transfers genes artificially from one organism to another.
- A **dominant allele** (capital letter) will always hide the appearance of a recessive allele (lower case letter).
- An **ecosystem** is made up of living things and the environment and interactions between them. Each ecosystem has its own variety of living things (**biodiversity**). Biodiversity is reduced by the impact of human activities, including using resources.
- There are competing priorities for the use of the **Earth's resources**, because people have different priorities.
- Examples of strategies for protecting planet Earth include: recycling materials, conservation programmes, protecting sensitive ecosystems, programmes to protect endangered species, managing ecosystems.
- Living things are linked through **food chains** e.g. wheat crop ➔ rabbit ➔ fox ➔ invertebrates ➔ microbes
- **Photosynthetic organisms** – mainly plants are always at the start of a food chain; animals are consumers because they eat ready-made food in the form of plants or other animals. Many food chains overlap, forming a food web.
- **Populations** are interdependent: the predator population follows that of the prey, with a time lag.
- **Carbon dioxide** is used by plants, during **photosynthesis**. It is returned to the atmosphere by living things when they carry out respiration, decay or when we burn fossil fuels or other once-living things, e.g. wood.

Atomic structure

- **Atomic number** = number of protons (and electrons). **Mass number** = number of protons + neutrons.
- Electrons are in **energy levels** (shells). The maximum number of electrons in each shell is 2.8.18.32
- Isotopes are atoms of an element with same number of protons but different numbers of neutrons.

Bonding

- Atoms lose, gain or share electrons to form full outer shells.
- Losing electrons forms positive ions and gaining electrons forms negative ions.
- Metals form ionic compounds (high melting point solids). Non-metals share electrons (covalent bond). Covalent compounds are mainly low boiling point liquids or gases.

Chemical reactions

- **Symbols** of elements and formulas of compounds are used in balanced, chemical equations to represent and quantify chemical reactions.
- **Atomic mass** enables us to calculate molecular mass, weights of reactants and products, percentage compositions, formulas of compounds etc.
- 1 mole of a substance is its molecular mass in grams. It always contains the **Avogadro number** (6.02×10^{23}) of molecules. 1 mole of any gas occupies $24dm^3$ at r.t.p.
- Acids produce H^+ ions in aqueous solution. Alkalis produce OH^- ions in aqueous solution.
- Neutralisation is the reaction: $H^+(aq) + OH^-(aq) \rightarrow H_2O(l)$
- Oxidation: gain of oxygen, loss of hydrogen or electrons.
- Reduction: loss of oxygen, gain of hydrogen or electrons.

Organic chemistry

- **Hydrocarbons** only contain hydrogen and carbon. The simplest are alkanes (C_nH_{2n+2}). These are **saturated** compounds – they only contain single bonds.
- **Alkenes** are hydrocarbons containing a **double bond** (**unsaturated** compounds) e.g. C_nH_{2n}.
- Saturated compounds undergo **substitution reactions**.
- Unsaturated compounds undergo **addition reactions.**
- Crude oil is separated by **fractional distillation.**
- Long-chain, less useful molecules are split up (cracking) into more useful short chain molecules.
- **Polymerisation** joins small molecules (monomers) into giant long chain molecules (polymers) e.g. n C_2H_4 (ethene) $\rightarrow$ $(C_2H_4)_n$ (polyethene/polythene).

Earth and atmosphere

- The reactivity series lists metals in order of reactivity.
- The **most reactive** metals have the most stable compounds so are very difficult to extract from ores. For example, **electrolysis** is needed to extract sodium.
- **Less reactive** metals have less stable compounds and can be extracted by **chemical reduction of the oxide**,e.g. zinc.
- Metals can be displaced from their salts or oxides by more reactive metals.
- Ammonia is manufactured by the Haber process:

$$N_2 + 3H_2 \xrightarrow[\text{Fe catalyst}]{\text{250 atm 550°C}} 2NH_3$$

- Ammonia is used to make nitrogenous fertilisers:

$$NH_3(aq) + HNO_3(aq) \rightarrow NH_4NO_3(aq)$$

- Artificial fertilisers can cause water pollution.
- Three rock types are: **igneous** (formed when molten magma solidifies), **sedimentary** (compacted layers), **metamorphic** (heat and pressure on other rock types)

The elements

- The periodic table lists elements by **atomic number**.
- Elements with same number of electrons in their outer shell are in the same vertical **group**.
- Horizontal rows are **periods**; moving left to right across a period sees a change from metals to non-metals.
- Transition metals have an **incomplete penultimate shell,** giving them characteristic properties, e.g. variable valency, coloured compounds, etc.
- **Alkali metals** lose their one outer electron: $M - e^- \rightarrow M^+$ (uni-positive ion). **Halogens** gain an electron to complete their outer shell: $X + e^- \rightarrow X^-$ (uni-negative ion).
- Noble gases have complete outer shells and are inert.

Measuring reactions

- Reaction occurs when particles collide with at least the activation energy (E_{Act}).
- Changing conditions, such as temperature, pressure, concentration, surface area, catalyst, alters the number of fruitful collisions, affecting the rate of reaction.
- Energy is given out (**exothermic**) or taken in (**endothermic**) whenever reactions take place as a result of breaking and making of bonds.
- **Bond energy**: the energy (+kJ/mol) needed to break a bond. The **heat of reaction** (ΔH) can be calculated from bond energies.
- Reversible reactions can proceed in either direction and reach **equilibrium** when forward rate = backward rate.
- Position of the equilibrium can be affected by changing the conditions. **Le Chatelier's principle** states that if a change is applied to a system in equilibrium the system will move to minimise the change.

Electricity and magnetism

- **Electric current** = flow of charge. **Voltage** = difference in energy carried by a charge between two points. **Resistance** = difficulty for current to go round a circuit.
- When a current-carrying wire is at **right angles** to a magnetic field there is a force on the wire.
- **Electromagnetic induction** needs a changing magnetic field near a wire or a wire moving in a magnetic field.
- A transformer needs a.c. to create a changing magnetic field to induce a voltage in the secondary coil.

voltage (V) = current (A) × resistance (Ω)

electrical power (W) = voltage (V) × current (A)

energy transferred (J) = power (W) × time (s)

energy transferred by an appliance (kWh) = power (kW) × time (h)

Forces and motion

- **Speed** tells us the rate at which an object moves.
- **Acceleration** tells us how velocity changes each second.
- When two bodies interact the forces they exert on each other are **equal and opposite**.
- If there are balanced forces on an object, it remains at rest or moves at a constant speed in a straight line.If the forces are **unbalanced** it will **accelerate**.
- A car's **braking distance** depends on: speed, braking force, car's and occupants' mass. **Stopping distance** depends on: speed, driver's reactions, braking distance, car's and road's condition.
- When objects fall, air resistance acts on them. The amount of air resistance depends on how fast the object is falling and its shape. The bigger the weight of the object, the faster it must fall to reach terminal velocity.

weight (N) = mass (kg) × gravitational field strength (N/kg)

$$W = mg$$

$$F = m a$$

Waves

- Waves transfer energy without transferring matter.
- **Transverse waves**: the oscillation is at right angles to the direction in which the energy travels.
- **Longitudinal waves**: the oscillation is in the same direction as the direction the energy is carried.
- The **electromagnetic spectrum** is a family of transverse waves that travel at the same speed in a vacuum.
- **Diffraction** occurs if waves pass through a gap narrower than the wave or past the edge of a solid barrier.
- Sound is carried by particles in a medium **vibrating**, so sound cannot travel through empty space. The **pitch** of a musical note depends on the **frequency** of the **vibration**.
- Earth has a **layered structure**. The thin outer layer is made of large plates. Where the plates move together mountains may fold, volcanoes erupt and earthquakes occur.

wave speed (m/s) = frequency (Hz) × wavelength (m)

$$v = f\lambda$$

Energy

- Energy is transferred when there is a temperature difference between two bodies.
- Metals are good conductors of energy. Their **free electrons** have more kinetic energy when the metal is hot and help to **transfer energy** to the cooler part.
- Much of the energy we use comes from non-renewable resources, e.g. fossil fuels, such as coal and natural gas.
- Renewable sources of energy are those that are continually replaced –like the Sun, wind and waves

power (watts) = rate of transfer of energy (joules / second)

work done (joules) = force (newtons) × distance (metres)

$$PE = m g h$$

$$KE = m v^2$$

Radioactivity

- Radiation ionises molecules in the material it passes through. Radiation can damage cells in the body.
- Background radiation comes from natural sources and man-made sources in the environment.
- The half-life of a radioisotope is the time taken for half the nuclei present to decay.
- **Alpha** (α) particles are made of 2 protons and 2 neutrons. **Beta** (β) particles are electrons. **Gamma** (γ) radiation is short wavelength electromagnetic radiation.

Earth and beyond

- The solar system consists of the Sun, nine planets, the asteroid belt and a number of comets. All the bodies in the solar system are held in orbit by gravity.
- The Sun is a star in the Milky Way Galaxy, which is one of many millions of galaxies in the Universe.
- Stars are formed when the force of gravity pulls clouds of dust and gas together.
- All the galaxies in the Universe are moving apart very quickly. The distant galaxies, with bigger red shift, are moving away faster than those nearer to us. This suggests that the Universe was formed by a big bang, which threw all the matter out in different directions.